"As academia faces a reckoning, in part of its own making, from the outside and from within, Hyland and Lewis offer a framework for a new poly-directional path forward. The authors extend various nodes—from Harney and Moten's 'undercommons' to Agamben's remaking of Western ontologies and Benjamin's philosophies of aesthetics—to redress the academy's perennial reluctance to pivot toward new ontological possibilities of (creative praxis as operative to) knowledge gathering, building, and redistribution(s). In the process of the internal squabbles, the doors have been left ajar for fascist tacticians to make gains in dismantling the entire project. The lingering remnants of a neoliberal hangover. Hyland and Lewis's undertaking is a disassembling of the existing institutional frameworks and its binding concepts of research/researcher contributions to productive non-production. The authors find purpose in scattering the pieces for a future reassembly from a studio-based alchemy—contributing new usage for an epistemological drift."

— Triton Mobley, University of Pittsburgh

"In *The Alchemy of Research-Creation*, Peter Hyland and Tyson Lewis question the Enlightenment's stark separation of art and science. They raise the possibility of an alternative way of approaching the apparent division of knowledge and pleasure—of research and creativity—by considering alchemy as a practice located in an understanding of space-time that assumes a continuity rather than a separation of knowing and creating. They imagine a different way of being in the world, one that sees the world as an always unrealized collection of possibilities in which transformation is not simply a means to an end but the definitive human activity. They provide not only grounded defense of the alchemical studio as a crucial space for a set of necessary and vital practices but also guidance on how this alchemical studio can become a space for education. They do not offer advice on method but, to use their language, sketch out what they, as teachers and artists, have done in the classroom as studio. This book is not so much an argument for alchemy as it is a work of alchemy, one that invites its readers to embrace the risk of inhabiting a world that is defined by possibility."

— James L. Kastely, University of Houston

"Lewis and Hyland are the new alchemists! A very important critique of positivist scientific knowledge cults and mystifying artists' practices. *The Alchemy of Research-Creation* reconsiders studio practices without recourse to neoliberal/technocapitalist 'interdisciplinary' imperatives or fascist drives toward achieving perfection. We are led down into a cryptic world of practices pasts and possible futures that invite us to undo the binarity of studio vs. lab, drifting alchemically away and within legitimacies and illegitimacies of knowledge that exist between and inside both. There is little interdisciplinarity here. These new alchemists pursue something more akin to an intra-disciplinarity that is far more fit for the complexity of twenty-first century research practices than what we currently have. The splits between head and hand, knowledge and pleasure, science and art are not made whole but rather are configured through five gates to different research in ways that matter."

— Annouchka Bayley, University of Cambridge

The Alchemy of
Research-Creation

The Alchemy of Research-Creation

On Pataphysical Experimentation, Dark Infrastructures, and Antifascist Studio Practices

Incomplete First Edition

Tyson E. Lewis and Peter B. Hyland

SUNY PRESS

Published by State University of New York Press, Albany

EU GPSR Authorised Representative:
Logos Europe, 9 rue Nicolas Poussin, 17000, La Rochelle, France
contact@logoseurope.eu

For information, contact State University of New York Press, Albany, NY
www.sunypress.edu

Library of Congress Cataloging-in-Publication Data

Names: Lewis, Tyson E. author | Hyland, Peter B. author
Title: The alchemy of research-creation : on pataphysical experimentation,
   dark infrastructures, and antifascist studio practices / Tyson E. Lewis
   and Peter B. Hyland.
Description: Albany : State University of New York Press, [2026] | Includes
   bibliographical references and index.
Identifiers: LCCN 2025041178 | ISBN 9798855806441 (hardcover) | ISBN
   9798855806465 (epub) | ISBN 9798855806472 (PDF)
Subjects: LCSH: Alchemy | Occultism and science | Pataphysics | Creation
   (Literary, artistic, etc.) | Arts--Research
Classification: LCC BF1439 .L48 2026
LC record available at https://lccn.loc.gov/2025041178

Contents

Acknowledgments

We would like to thank a motley crew of misfits for their wayward assistance with the wrangling of this beast of a book. First and foremost, early readers such as Chris Moffett should be commemorated for their sacrifices. We can say that Chris helped transform an already baroque book into an even more labyrinthian puzzle. Feel free to send your complaints to him directly (written in invisible ink, of course). Our graduate student assistants who toiled over the manuscript, helping hunt down obscure references, gather image permissions, construct archaic indexes, and so forth should also be noted: Jillian Wendel, Shahrzad Abbasi, and Nasrin Tork. They were very gracious, diligent, and never once asked, "What the f*ck is this f*cking book anyway?" Conversation with Erin Manning, whose support for this project has been invaluable, and the work of contributors such as Andrew Culp to the dark infrastructure special issue of *Techniques* (cocurated by Tyson and Nasrin) were also extremely helpful in thinking through the concepts of darkness and drift. We owe great gratitude as well to Kim Willis, conjurer extraordinaire at the Jo Ann (Jody) and Dr. Charles O. Onstead Institute for Education in the Visual Arts and Design at the University of North Texas (UNT), for her vital contributions to our project, in relation to this publication as well as the related biennial Education as Experimentation symposia that are a part of the life cycle of our ongoing work. Additionally, publication of this book would have been impossible without the support of mages holding leadership roles in the UNT College of Visual Arts and Design and Department of Art Education. Likewise, the Onstead Foundation and C. Loren Vandiver deserve our thanks for the sorcery and enchantments they provided to wrest this book into this world and others beyond it. There are others to note, but they shall remain in the dark. . . .

AWKWoRD

Erin Manning

Be genuinely experimental, balancing formal constraints, open procedure, and chance and fortune. The idea here is not to confirm results based on an existing hypothesis but rather see what happens when a protocol is collectively performed.

— Lewis and Hyland 2022: 72]ta

The dark infrastructure of the studio (rules, windows, fireplaces, and so forth) incubates awkward drifts between these two. What the laboratory attempts to stabilize through the empire of algorithms, the studio intentionally sets adrift.

— Lewis and Hyland 2025

Through our crypto-genealogy of studioing practices we discover three very specific modes of risk: unprofessional risk, awkward risk, and finally, at its most extreme, the risk of madness.

— Lewis and Hyland 2025

The Alchemy of Research-Creation: On Pataphysical Experimentation, Dark Infrastructures, and Antifascist Studio Practices is a proposition and a pact. In its pataphysical logic, its pact is to awkwardly draw us in to the architecting of an adjacent spacetime, to drift us into alchemical act, obliquely. In the same gesture as it makes a proposition: experiment. Don't manage what has already founded a legacy, grow it. Live its paradox and learn from the movements it emboldens. Idle quickly! Slow up! Go with it!

Going with it, obliquely, makes its own demand. Decenter your-self and follow time's bend. Make your coming into thought a practice of shifting the conditions of the time of learning and the space of thought. "Suspend educational processes, subject positions," write Lewis and Hyland, in their protocol-based proposition for another kind of learning into and across the virtual realm (2022:

1). Don't reduce it to something you know how to know. Learn it otherwise, in the study of what is made possible from the vantage of awkward timespace. Be made by the pataways, by the pataprotocols it calls forth. Learn to bend into new shapes. Lean into the discomfort. Live the awkwardness of the puzzle of a direction-without-direction, of an alchemical now where "one is always already in a social situation in which one cannot fully escape. One is always dealing with others, wrapped up with others . . . exposed to the possibility of taking a wrong turn or making a mistake that violates certain social laws, norms and expectations" (2025).

I begin in the middle because that's where the alchemy does its most potent work, and where Lewis and Hyland take us most awkwardly. In this space of non-space (like the relation of non-relation), a particular kind of study takes shape. Because a space of non-space can only ever be the aliveness, the awkwardness of the sociality of space-*ing*, of *studioing*. Nothing is presupposed here—in the awkwardness of the sociality it calls forth, the coming into itself of studioing is in no way separable from the how of its coming into the shape it takes. This studioing, a study in the spacing, moves at the pace of a research-creation of radical awkwardness where the constraints must be crafted each time to enable the conditions for its dis-orientation.

When research-creation disorients, it has begun to do the work of crafting worlds. "The alchemical pataphysics of research-creation can act as a counter to the capitalist sorcery of control and the spacetime of the studio (in all its pliable manifestations) can safeguard an undercommons full of unprofessional, awkward, and mad swerves" (2025).

It might be nice to believe that the studio is the *site* for this practice, and that once properly sited, a particular kind of back-and-forth that gives credo to both research and creation—separate and overlapping—might occur. Lewis and Hyland are not having it: Research-creation, in its bubbling awkward sitedness, crafts only obliquely, in the pata of logic's refusal of the binary, and of the hyphen that would stabilize it.

In *The Alchemy of Research-Creation*, there is an abiding concern about space, a worry that the space of making may become too generalized in the whimsical accounts of all that research-creation can do. They are right to be concerned. When research-creation loses its pragmatic tether to the speculative dimension that gives it life, it is too often because it has been captured by the regime that makes the MFA the new MBA. Give it to the market and it will produce the necessary offspring—those interdisciplinary cross-appointments; the subgenres that never stop tethering research to the social sciences, art wrapped up like a perfect department store gift. All within the logic of the colonial-capitalist institution.

The alchemical transformation, the studioing, threatens to sidle research-creation out of its legibility. If it sites it, it does so at the risk of cauldrons exploding and experimenters losing their way in the distraction of all that uselessly bubbles up.

In the mix, the ingredients come from far and wide. Education is suspended in lieu of something more wild, more queer, more slanted. It's hard to enclose it, to set its parameters. The studio keeps shape-shifting, the alchemy as much about spacetime itself as what it cooks up.

As with their earlier *Studious Drift*, the suspension of educational processes requires the suspension of the self. Self is not where the creativity bubbles up. Learning occurs in the sociality of a fabulation that exceeds any one account where it comes from. Fabulation here, power of the false, is what grows askance in the excess-over myth, the excess-over the last word that would say it like it is. The emergent sociality that ensues studios, which is to say that it alchemically reconcocts the terms of the agreement that would otherwise claim to know.

Fabulation requires its intercessors: Its stories grow from the middle thanks to all that intervenes. These are interrupted stories, not-quite-true accounts laden with exaggeration. Trickster-tales. The oblique is felt through them, and these are the paths Lewis and Hyland beckon us to follow, their accursed share the seductive force of what they leave behind.

This might be said to be the *problem* of *The Alchemy of Research-Creation*: How to be sure to refrain from claiming that accursed share; how to refrain from capture? The experimentation this requires is not unlike juggling with a cow head, a lemon, two candlesticks, and an idle thought. If the anarchic share is what keeps the archive from closing in on itself, if this speculative edge is what stops research-creation from mortgaging itself to the academic institution (and its affiliate, the art market), if it is what keeps it in the abeyance of the Self (which is to say whiteness, neurotypicality), if what exceeds the shape things take is vital to *stay with the problem*, how to study it without enclosing it? How to write it?

The book bears the scars of this question. If it is read in paper form, its pages will swell with each rereading, having taken up the steam of the bath of thoughts unparsed. The pataphysical force will infuse the pages with the moistness of all that leaks between. The studioing has to read itself into being, has to trouble reading.

Lewis and Hyland write: "The following book offers a spatial-temporal *retrieval* of the vestigial potentiality of the alchemical studio as providing the hospitable conditions for experimenting with research-creation that does not fall into the trap of (a) merely submitting the arts to the reign of the sciences or (b) hybridization. This is a dark spacetime that drifts in the zone of contact between art and science, studio and laboratory, head and hand, pleasure and knowledge" (2025).

The trap: that this will all come together too elegantly; that the oblique will be straightened; that its shape will reveal itself as equal to the problem.

Studioing is a topological shape-shifter, which commits only to taking up the space of the time it creates. Research-creation hyphenates its discomfort with being either-or. This is not knowledge-creation in any standard sense. Knowledge de-standards here, its normopathy enlisted to other ends. That is to say, the normopathy will lurk, aiming to whiten the frayed edges so they can be recalculated, counted back in. But the awkward thickening of the pages will make it difficult to seam it all together anew.

The dance will continue until another logic emerges. This will never happen once and for all: a paraontology (which will here pronounce itself pataontology) will craft itself. But it will be fragile, as all experiments are. And it will fail, blowing up the studio, more than once.

Pataontologies are the meta of meta, the beyond-sense of alchemical reorientations. They don't only unsettle, they re-geography.

In the blowing up, it's not so much that a new site is built than that a new rhythm is called for: patarule. For Lewis and Hyland, darkness emerges here. Against the transparency of rule-bound systems, an adjacency shape-shifts into existence. I wonder here about the opacity of a poetics of relation, à la Edouard Glissant, that is not dark or shadowy so much as it is simply not any sum of any parts. Consent not to be a single being. Opacity as relation of non-relation does not so much refuse and it gathers in. More-than. This could be blinding.

Blackness, after all, per du Bois, *is a problem for thought*. The opacity is not the shade colonialism gives it. It is the poetics that exceeds it.

Let this "an-aesthetic supplement" take you there, into that excess. Feel into all that anarchives itself into the drift of a form of study that studios you. Let the alchemy do its work, awkwardly, in the blackness of a surround that is always more-than the shape things take. Be proliferated. Let it rip.

Just don't expect transparency.

Initiation

The Modern World (Divided)

In the eighteenth century, science and art started to part ways, parceling out two divergent approaches to understanding reality. This epistemological schism was accompanied by different practices of inquiry, different forms of attentiveness, different kinds of problems, and different grounding assumptions concerning knowledge. For its part, art became increasingly wedded to the aesthetic, taste, imagination, and subjectivity (questions of the perception of nature), while science embraced objectivity, falsifiability, and reproducibility (questions concerning nature as such, in itself). As Isabelle Stengers points out, we can see this bifurcation announce itself clearly in a key shift in the writings of Galileo. Galileo's early texts were written in the form of dialogues, featuring quasi-fictional characters that embody different scientific theories. Critically, all these perspectives were fictions, and the debate concerned *aesthetic* or *rhetorical* techniques for convincing the reader of the superiority of one position over another. Yet, with the invention of the inclined plane, all stories about falling bodies became subjected to an *objective* and *observable* test. Galileo's "experimental apparatus" allowed the author to "withdraw" and let the "motion *testify* in his place."[1] Galileo thus produces a gap between fact and fiction through which science can differentiate itself from everything that is not science (meaning not empirically and practically demonstrable). Dialogic literature and its aesthetic conventions gave way to proofs and demonstrations in the natural sciences. Subsequently, Alexander Gottlieb Baumgarten inaugurated the modern field of "aesthetics" as separate from science and its concern for physical facts. Whereas Galileo's science emphasized the object experienced, Baumgarten's aesthetics emphasized the experiencing subject. Thus, at the very heart of modernity lies a series of distinctions between science and art, subject and object, fact and fiction, literature and proof. Take for instance Lorraine Daston and Peter Galison's history of Western "mechanical" objectivity, which gained a certain moral high ground by partitioning itself off from the dangers of subjectivity and aesthetic judgement.[2] Negatively, mechanical objectivity separated itself from the presumed temptations of the sensual and the pleasurable, and positively it aligned itself with self-discipline, self-restraint, and honesty. Daston and Galison summarize, "It is a vision of scientific work that glorifies the plodding reliability of the bourgeois rather than the moody brilliance of the genius [artist]."[3] Pragmatically speaking, in the late nineteenth century, this meant that faith was increasingly placed on the impersonal machine to make scientific observations and take scientific measurements. The machine enabled the head to be separated from the hand so that frailty of the human senses would be minimized while reason and restraint could be maximized. Case in point is William Anderson's 1885 introductory address to the Medical and Physical Society of St. Thomas's Hospital in which he argued that "scientific understanding had not only made artistic insight supererogatory, it had shown that the artist could prove to be a liability"[4] for scientific objectivity (and thus scientific advancement).

In a slightly different context, Giorgio Agamben observes that this division of art and science was accompanied by a hierarchical separation of the senses, with sight and hearing taking on increasingly privileged positions as scientific and objective senses over and against the more debased sense of taste (or more broadly, sensorial pleasures, including those of taste, touch, smell, and so forth).[5] Within the Cartesian mapping of the body in relation to the world, one sense knows without enjoying, and the other enjoys without knowing. This hierarchical division of labor *within* the body separated the nobility of human reason from the animality of the senses, effectively cleaving the human from its own sensorial grasp of the world. Looking for alternatives, Agamben then turns to ancient Greece and highlights the unique role of *divinatory science*, which occupied an aporetic position between science and art. He also points to philosophy as a form of divination predicated on the existence of the *daimon*—or occult presence necessary for philosophical inspiration. Yet, in the modern period, the peculiar position of divinatory science as existing in a zone of contact between art and science gave way to the natural sciences (knowledge of what is known) and aesthetics (pleasure in what is not known, for which there is no adequate concept). Lost was an understanding of the human as *homo sapiens* or the being with a *taste* for knowledge.

These institutional and embodied divisions cannot be fully understood outside the context of the broader sixteenth-century capitalist revolution. In 1864, Karl Marx gave the following critical and utopian view of social struggles against capitalist exploitation:

> We speak of the co-operative movement, especially the co-operative factories raised by the unassisted efforts of a few *bold* "hands." The value of these great social experiments cannot be overrated. By deed, *instead of by argument*, they have shown that production on a large cale, and in accord with the behests of modern science, may be carried on without the existence of a *class of masters employing a class of hands*; that to bear fruit, the means of labor need not be monopolized as a means of dominion over, and of extortion against, the laboring man himself; and that, like slave labor, like serf labor, hired labor is but a transitory and inferior form, destined to disappear *before associated labor plying its toil with a willing hand, a ready mind, and a joyous heart.*[6]

Capitalism is dependent on the split between head and hand, between thinking and making. The science of capitalism is predicated on a fundamental bifurcation of the body (embodied alienation) that must be overcome by a revolution that is equal parts economic, political, and aesthetic. Socialism, as depicted in this quote, is a reconfiguration of the human body and its capacities to make (with willing hands), think (with a ready mind), and feel (with a joyous heart). This would amount to a knowing through doing (deeds instead of arguments) and a doing that is also a knowing. For Marx, socialism would be a way of knowing through the hands that would no longer demand a managerial class as disembodied head. Surprisingly, the science of capitalism offered up by Marx is also and equally an aesthetic critique of an impoverished body alienated from itself.

Further developing Marx's insights, Alfred Sohn-Rethel argues modern science and the experimental model developed in the laboratory were both essential outgrowths of the fundamental capitalist division of labor between the head and the hand, between thinking and making, between judgment and production.[7] Capitalism cleaved apart mental and manual labor, which were once combined in feudal and medieval artisanal economies. Sohn-Rethel traces the intellectual powers of conceptual reasoning back to historical origins in capitalist commodity production, revealing how the sciences were largely produced out of and developed in response to economic pressures. The sciences are, according to this argument, the socially necessary forms of thinking in a capitalist epoch that demands a division of labor between the head of the capitalist and the hands of the workers, a high level of rationalization, management, and automation of manual operations, the privileging of the logic of abstract markets and of mechanistic thinking over situated, contextually specific handicraft making, and a reification of value (subjective and aesthetic) into a single, universal, and formal function (money). Descartes's separation of mind and body as well as the purified, Kantian transcendental subject both became useful "fetish concept[s]" that supported a science that was "equipped for its socially necessary tasks, but only with false self-awareness."[8] Galileo's and Newton's concepts of inertial motion, for instance, are abstract, mathematical, intellectualized models of nature separated from manual labor and the experience gained from working with one's hands. In other words, philosophy was able to provide the transcendental arguments necessary to legitimize the fundamental separation between minds and bodies, thinking and feeling, so essential to the modern bourgeois sciences.

The divisions of labor between art and science, knowledge and pleasure, head and hand were given concrete, spatial form in terms of the hierarchical division between the studio and the laboratory, each with its own claims to "experimentation." Schematically put, during the rise of the modern, Western period, the laboratory became the space for testing the verifiability of a hypothesis within a framework of a specific understanding of nature and knowledge. In the laboratory, the experiment was safeguarded from the waywardness of the human hand. Various instruments and measuring devices were used to carry out the experiment, separating direct contact between the scientist and the phenomenon. The scientist interacted with the phenomenon *indirectly* through reading instruments that produced quantitative measures. Thus, the laboratory became a space in which fallible, human sensation was replaced by mechanically mediated numeration to ensure objectivity. Knowledge was separated from sensation and science from aesthetics through laboratory practice and laboratory infrastructure. On the other hand, the studio became the space for the creation of new (sometimes exceptional) sensations. The intimacy between the subject and the object was emphasized (rather than downplayed). Indeed, the object was only authentic if it demonstrated the mark of the maker as an expression of individual genius. The myth of the modern, Western studio as an extension of the body and spirit of the artist was born, ending with the high modernist fetishization of the celebrity artist and his/her celebrity studio.[9] Opposed to the laboratory, the studio was conceptualized in qualitative, sensorial terms (the less mediation between subject and object, the better). While this separation enabled each domain to develop specialized kinds of knowing and perceiving the world, the separation has also led to an increasing sense that something has been lost.

How to describe the space that separates and conjoins the head and the hand, knowledge and pleasure, science and art, studio and laboratory? This location—not reducible to either side of the equation—must be continually disavowed in order for the division to maintain itself. As Bruno Latour argues, the first Enlightenment was predicated on a purification through the division of the world (between art and science, head and hand, aesthetics and knowledge). He writes, "The obscurity of the olden days [the premodern period], which illegitimately blended together social needs and natural reality, meanings and mechanisms, signs and things, gave way to a luminous dawn that cleanly separated material causality from human fantasy. The natural sciences at last defined what Nature was, and each new emerging scientific discipline was experienced as a total revolution by means of which it was finally liberated from its prescientific past, from its Old Regime."[10] But, as Latour also points out, the abject position betwixt and between these divisions *haunts* the system from the inside out as a profane excess that can only be glimpsed obliquely, out of the corner of an eye when one is not paying attention to what one ought to be paying attention to (according to modernity's self-proclaimed principle of purification). This questionable no-man's-land, which prefers not to abide by divisions and instead multiplies through mixing, is a space for experimentation. If we follow Latour and assert that "everything happens in the middle. . .everything happens by way of mediation, translation, and networks" then we also have to admit that "this space does not exist, it has no place" and is "unthinkable, the unconscious of the moderns."[11] Whomever wanders into the weird and uncanny space that divides the artistic studio and scientific laboratory does not play by the modern rules, and is sure to bump into unwanted monsters, critters, wonderous beings/ becomings, all the while risking their reputation, respectability, recognizability, and the very intelligibility of their statements and enterprises. Yet perhaps it is time to *consciously* and *overtly* take such risks and once again experiment with what lies in the shadow of the modern world as an untapped, wayward potentiality. Perhaps it is time to openly embrace the unthinkable space that, as noted by Latour, must be continually denied by the divided Western world even as its (very effective!) myth of purification is only possible because of the multiplication of such unthinkable monsters.

Open acknowledgment of and experimentation with the denied space of modernity is one way of describing research-creation. Although there are a variety of definitions one could offer as well as various genealogies and stages through which research-creation can be tracked, perhaps the most concise definition of research-creation has been given by art historian Claire Bishop. At its best, "artistic research can push against the limits of academic research in two ways: first, by allowing personal narrative and challenging an objective relationship to truth via fiction and fabulation (a tendency already present in academia via feminism and black studies); and second, by presenting research in aesthetic forms that exceed the merely informative (the pleasure of a well-crafted story; connections and juxtapositions that surprise and delight)."[12] Notice that for Bishop, research-creation can cut across divisions separating fiction from fact through various fabulations, and that these fabulations further complexify easy distinctions between knowledge and pleasure. Natalie Loveless also calls for a new form of research-creation as one (minor but important) intervention into the divisions that structure knowledge production. For Loveless, research-as-creation "produces hybrid forms, defamiliarizing and uncanny, that oscillate between more than one 'species' of production and that result in 'offspring' that are often unthinkable, illegible with in the current institutional frameworks of academia."[13] No longer separated by a hierarchically inflected divide—that inevitably privileges science over art, research over creation, head over hand—research-creation is an unpredictable contamination making each unfamiliar or unrecognizable to itself (and to its purported other). It is promiscuously polyamorous in terms of theory and practice, preferring not to delimit the drift of desire that does not know how to abide by disciplinary distinctions.

While agreeing with the general capacity of research-creation to trouble the divisions that structure the modern, Western world (and especially the academic division of labor), we are surprised by a lack of reflection on the *space* of such an endeavor, or rather how spatial problematics are often replaced by questions of time. Stated differently, the discourses theorizing research-creation do not rigorously address the issue outlined above—the division between the artistic studio and the scientific laboratory—and as such leave this space "unthinkable." Take for instance the important work of Bishop. Bishop worries that one of the central traps of research-creation is a mere reproduction of the information-saturated space of the internet, which produces an increasingly taxing experience of perpetual searching (rather than researching), of being lost in a sea of information without any possibility of orientation, interpretation, or synthesis. While this strategy might have been an effective sabotage of white, male, academic authority in the '90s, Bishop highlights its problematic nature in a post-truth era characterized by increasingly short attention spans and conspiratorial rabbit holes. In response to this trap, Bishop points to artists who have overcorrected, resulting in a return to a singular, authoritative voice speaking from the space of the forensic laboratory. Although capable of cutting through the whizzing and buzzing noise of the first position, this option seems to fall back on didacticism and even authoritarianism, demanding the erasure of precisely the ambiguities that research-creation offers as an intervention into the world divided. Interestingly, when it comes to rehabilitating the unique capacities of research-creation, Bishop fails to address the issue of postdigital space. Indeed, Bishop cites the work of Egyptian artist Anna Boghiguian as someone who successfully avoids the two extremes previously outlined, but in her discussion of Boghiguian's work, Bishop shifts from spatial analysis (coding the space of research-creation as the digital sphere of the internet or the science of the forensic laboratory) to history and, more generally, temporality. Boghiguian's creative and critical, informed and passionate, synthetic and open-ended, intellectual and embodied, installations concern "time spent" on various sites where important events in history have taken place. As such, spatial questions seem to fade into the background for Bishop.

Although Loveless proposes a radical politics associated with research-creation at the *end of the world* (which she characterizes as the ominous sense of democratic and ecological disasters await us in the near future as neoliberal capitalism stands largely unchallenged), she likewise tends to ignore the need for a concerted effort to think through the spatial problematics of this end. For instance, Loveless rightly calls for experimentation with "multimodal ways of working in and with the arts and humanities attuned to more-than-human social justice" and the "importance of focusing on *how we do what we do* within university spaces as we face the end of the world (as we know it)."[14] Although an intervention into university spaces, the nature of this spatial occupation is not addressed. She focuses instead on the "*temporal* and *material* attunement" of the *how* of her practice. Even if we think of space in terms of the how by asking what kinds of spaces are generated through research-creation (rather than preexisting such practices), Loveless gives little guidance. For us, a more exacting examination of the space of research-creation is needed in order to address this gap. While Loveless seems to reject the idea of the laboratory, at the same time, we could convincingly position her theory of research-creation within a "post-studio" understanding of educational space,[15] which rejects the studio as a white cube created for art world consumption by predominantly white men.[16] Research-creation has a time and a practice but seems to exist nowhere—floating in some kind of limbo between the artistic studio and the scientific laboratory. What then is the spatial dimension of Loveless's "how"? Stated differently, how might we rethink the spatialization of this how, not in order to then exclude the temporal, but rather in order to build a more dynamic spatio-temporal notion of research-creation—one in which space is not simply swallowed up by time but rather a constitutive dimension of practice?

In this sense, we find Matthew Fuller and Eyal Weizman's invocation of "investigative aesthetics" more persuasive than Loveless's research-creation precisely because of its spatial emphasis. Fuller and Weizman's investigative aesthetics proliferates differences between the studio and the lab while also exploring how "forking" between the two can "produce novel syntheses" that are "uncannily productive" of new ways of seeing and thinking about the world that contaminate the space of the lab with aesthetic concerns and populate the studio with tools, procedures, and methods of research once strictly bound to the lab.[17] Yet we are somewhat troubled by the idea of "novel syntheses" or mediations as a way of conceptualizing spatial and temporal dimensions of investigative aesthetics. Given the dominant status of the sciences in contemporary Western society, it is a plausible worry that the epistemology of the laboratory (verifiability) would dominate any attempted synthesis. Indeed, Fuller and Weizman's own practice is a case in point, as the fundamental notion of scientific investigation remains largely unaffected by the synthesis. Thus, spatial overlap turns out to be a hidden form of epistemological imposition of the laboratory over the studio.

Hannah Star Rogers argues that Art, Science, and Technology Studies (ASTS) focuses on practices in order to recognize similarities across traditional divisions between art and science, thus revealing how art and science are actually unstable categories. This is an interesting approach, with many implications for how we think of the space and time of art-science (or research-creation). For instance, Rogers is concerned with what happens when bioartists are transplanted into functioning biotech labs and how they are initiated into lab culture, rhetoric, and practices. In such situations, emphasis is placed on learning technical skills and knowledge in order both to use laboratory equipment while also critiquing science practices. Rogers points out how many artists that take up residencies in such laboratory settings avoid either reducing their practices to serving science (providing illustrations for scientific discoveries) or attempting to hybridize art and science. Instead, artists either insist on maintaining the integrity of the arts disciplines or use their practice to critique the sciences (and thus raise public awareness of scientific problems). But what is important for us is how Rogers describes the relationship between the studio and the laboratory in this process. The physical space of SymbioticA (an artistic research lab dedicated to bioart projects) was originally organized into a studio space and a laboratory space. The studio space was meant for the artists, but its design was predicated on certain assumptions about art making that were not very conducive to making biological art and as such, the studio was quickly abandoned and turned into an office and meetings space, with the bioartists opting instead to work in the laboratory, which contained tissue culture technologies. The studio was simply abandoned in order to enter into the laboratory in order to learn its processes and procedures in order to pirate its material and immaterial codes for critical ends. As one researcher whom Rogers interviewed stated, "they [the artists] are just another lab group."[18] In other words, the lab and its practices subsumed the artists. While the particular *practices* of science and its *technologies* embedded in the ethos of the laboratory might be critiqued by these artists, the artists did not fundamentally question the spatial-temporal logic of the laboratory or of science *as such*. The dependency of bioartists on scientific knowledge, practices, equipment, and spaces makes one question how critical their bioart can be, and how far it can challenge the regime of the sciences and the primacy of the epistemology of the laboratory.

Erin Manning's experiments in SenseLab—an event-based, collaborative project that was founded in 2003 at Concordia University, Montreal—presents one more variation of this problem. Manning foregrounds the need to think through the spatial and temporal problematics of research-creation, providing detailed analyses of the "spaZes" of SenseLab. She writes, "When I proposed SenseLab in 2003, my hope was to create an environment that would learn, over time, how to create conditions for new forms of collaboration across art, philosophy, and the political."[19] Underlying SenseLab is a sensitivity toward *hospitable* spatio-temporal conditions for research-creation. This is not a fixed notion of space but a space in motion, and thus a space that is inseparable from questions of temporal unfolding, becoming, and experimental events in collective emergence. The spelling of spaZe comes from an emphasis on "rez-zonance, reverberation, detour, dissensus" of a space that is never settled (a zone of thresholds).[20] We strongly identify with Manning's description of the SenseLab as an "activator of the more-than-human" that is "capable of breaking the container of the classroom,"[21] which is defined in terms of containing thought and bodies to neurotypical, white standards. While Loveless does not explicate questions of space and time, Manning integrates them into her analysis of research-creation practices, and whereas Fuller and Weizman fall back on doubtful claims of synthesis, Manning generatively focuses on immediacies/contact points (rather than mediations between science, art, and philosophy). And unlike the artists described in Rogers's ethnography, Manning's collective projects cannot easily align themselves with the spatial-temporal codes of the scientific laboratory. The space-time of SenseLab offers a minimal set of spatial coordinates necessary to activate alternative spatio-temporal folds and catalyze alternative forms of relationality between human and more-than-human bodies through complex compositions of materials and practices. Yet there is a terminological vacillation in Manning's writings that causes her to shift restlessly between describing such a spatio-temporal fold sometimes as a laboratory and sometimes as a studio. For instance, Manning clearly aligns research-creation in and through SenseLab with the concept of "black study"[22] and yet does not mention the connections between study (practices) and the study/studio (which embodies and embeds study in space-time). And in other publications, Manning refers to her own study practices as taking place in the semiautonomous zone of the studio (rather than the laboratory).[23] This vacillation is not merely a terminological problem as the question of the studio and the laboratory is an index of historical structures and practices of division that remain unresolved, and, in turn, could point toward another, more ambiguous (unthinkable) space-time that does not play by the laws underlying the initial split between art and science, pleasure and knowledge, hand and head that created these problems in the first place.

Our current book takes up the question of the space-time of these multiple experiments in reconfiguring science and art practices.[24] To do so, we wish to return to a lost spatial configuration, one that predates the split outlined above and remains a virtual possibility within the present, latent within the concept of research-creation (and its offshoots). We are talking about the medieval and early modern alchemical studio as a space-time machine copious enough to hold within itself a certain approach to the excessive pleasures of (non)knowledge lost with the modern reification of science and art and their respective compartmentalization in the laboratory and the studio. Through this repotentialization of a lost sense of alchemical studio space-time, we hope to avoid what we perceive as the major challenges facing research-creation. For instance, there is the concern that any attempt to forge together the disciplines that have codified the domains of art and science will inevitably favor the sciences precisely because of their dominance in the university and their status as producers of "valid" knowledge—especially within capitalism and its algorithmically enhanced, neoliberal forms. As David Rousell points out in relation to recent attempts to forge a new configuration or alliance between research-creation and "post-qualitative research," "It potentially leads to blended and diluted hybridizations that are not quite art, not quite philosophy, and not quite social science."[25] The worry expressed by Rousell is that mediations between art and science (and their respective space-time configurations) might avoid the limitations of instrumentalizing the arts, but in doing so yield unintended costs that water down both. Another upshot of this same phenomenon might be that the arts can be manipulated so as to produce a scientific aesthetic running cover for ideology (as when environmentally dangerous chemicals are conveniently greenwashed through aesthetic representations). In this sense, the arts become mere handmaidens to a scientific agenda or a ploy of capitalist extraction. Here we can think of Bishop's criticism of forensic approaches to research-creation, which merely reiterate the authoritative voice of the sciences and thus shy away from the more ambiguous pleasures of the arts, but also the foreclosure of the studio space by Rogers's bioartists who became dependent on the very laboratory practices and culture they hoped to critique.

On our reading, the problem is the starting point: a split that must be reconciled through the creation of a *relation* between separate disciplines, practices, and discourses. It seems that one starts from the existing poles of science and art in order move toward a mediated center, therefore taking for granted the legitimacy of the separation to begin with. Yet for us, there is another option, or what Agamben refers to as "contact." For Agamben, contact happens when there is a "deactivation and inoperativity" within a representation such that two entities "dwell in non-relation."[26] Contact undoes the premise of identity, and by extension, relationality. In our case, when science and art, knowledge and pleasure, and head and hand enter into contact with one another, their respective predicates are neutralized. Neither remembers what it is supposed to be or what it is supposed to do or which rules it is supposed to abide by. In this way, contact is neither an act of separation nor coincidence. Instead, it is an inoperative moment in which something (a way of knowing or feeling in the world) drifts away from itself and toward something else. As we will explore below, the alchemical studio is precisely the space and time in which contact can happen between affects, forces, concepts, and practices that would be unthinkable. It is a space-time of collisions, queer practices, and contaminated discourses, all of which make possible new uses within a world dominated by old separations (between head and hand, science and art, knowledge and pleasure). From the modern perspective, the alchemical studio might very well be dismissed as a premodern monster, a "horrible mismash of things and humans, of objects and signs"[27] needing to be dismissed. But we want to approach it as a site that still has a (vestigial) potentiality to offer contemporary research-creation as a space-time for contact between that which would otherwise be artificially divided and then relationally hybridized.

This shift away from a hybridized laboratory-artistic studio to an alchemical studio also means a fundamental rerouting of what research-creation means, aligning it more with the wyrd, queer, pataphysical, and neurodiverse forms of non-knowledge produced through occult practices than with the epistemologies of science or the aesthetics of art. Thus, a new (dark) dimension within research-creation can be opened up and experimented with that suspends the fundamental coordinates of the divided world.

Alchemical Studioing

As we have already explored in our previous book *Studious Drift: Movements and Protocols for a Postdigital Education*, the alchemical studio offers a paradigm of the studio as a space-time that prefers not to abide by the splitting of creation and research, pleasure and knowledge, hand and head that resulted from capitalist production, that were justified by transcendental philosophy, and that were embodied (perhaps unsuccessfully) in Galileo's scientific revolution. In this section, we will outline various dimensions of alchemy that are important for our argument before turning to an analysis of how these dimensions are spatially and temporally materialized in the alchemical studio.

First and foremost, it is important to note that alchemy is part of what is referred to as Western esotericism and its related occult practices. Although somewhat reductive (as the concept "Western" projects a set of artificial boundaries and divisions on a tradition that is defined by drift), Antoine Faivre's basic definition of esotericism is a useful heuristic for this study. According to Faivre, Western esotericism can be defined as a "form of thought" that includes an emphasis on correspondences (rather than causal analysis), living nature, imagination, and transmutation.[28] The concept of the occult can be traced back to Greek antiquity where there was a distinction drawn between observable phenomena that could be explained according to the four known elements and hidden, dark forces, energies, or "unspeakable qualities" that were simultaneously unexplainable and somewhat unnatural. This term was the most likely origins of the medieval concept of *qualitates occultae* or occult qualities.[29] For us, alchemy is an esoteric form of thought that attempts to contact these occult, dark forces through specific modes of transmutation.

According to Lawrence Principe, there are several consistent features of alchemy that can be found at its inception, including an a-disciplinary approach to inquiry in which science and art exist in a zone of contact that would trouble today's modern sensibilities. In the third century AD, the Greco-Egyptian alchemist Zosimos of Panopolis penned many texts in which science and art, theoretical and theological principles, speculation and experimentation entered into contact with one another, or rather, were *never out of contact with each other*. While it is common to reduce alchemy down to nothing more than the (futile) search for a recipe to transmute base metals into gold and silver, for Zosimos, this process was itself a philosophical exploration of ontological questions concerning the nature of substance (and of *monistic* ontology, in particular). As such, alchemy was an experimental philosophy that conjoined creation and observation, thinking and doing, head and hand, and immaterial and material dimensions of existence. As Principe writes, Zosimos, "had no reason to isolate his philosophical or theological commitments into special categories separated from the balance of this thought."[30] We can also cite the famous German physician and alchemist Paracelsus, who emphasized that alchemical work was a creative act and thus distinct from chemistry. Paracelsus wrote that alchemy "changes the character of simple bodies, and raises them up into higher states of existence. To exercise this power, not merely mechanical labor, but artistic skill, is required."[31] Mechanical labor such as steady observation and calculation had to be coupled with artistic skill, including aesthetic judgment, discrimination, and sensitivity toward particulars. For Zosimos and Paracelsus, alchemy was not simply a description of reality. Instead, it was an artistic science that altered reality through creative labors on and through material substances.

As a major literary form in the sixteenth century, the *books of secrets* tradition likewise recorded a host of methodologically rigorous, scientific practices in a highly codified, allegorical language of alchemical symbols and images that stood in stark contrast to Galileo's separation of creative, subjective exposition and objective description.[32] Unlike modern science and its interest in generalizable laws, books of secrets were more or less concerned with the marvelous, miraculous, and the exceptional (*meravilia*). They offered catalogs of the irregular rather than the regular functioning of nature. For instance, Giambattista Della Porta's book *Magia Naturalis* (1589) was a book of natural magic for investigating the extraordinary. The existence of the extraordinary suggested that the world was full of hidden correspondences that enabled analogical contact points between otherwise unrelated phenomena. The concoction of recipes and riddles in these books of secrets produced constellations that suggested hidden connections and *discordia concors* (harmonious discord) that could only be deciphered through equal parts scholastic interpretation, spiritual visualization, and experimental practice. Books of secrets were, in other words, a-disciplinary texts that proliferated in the space between art, science, and magic that is now, in the modern, Western world, occupied by a vacuum.

While modern, Western science increasingly became skeptical of the reliability of the senses (replacing the body with mechanical prostheses), alchemy was a corporeal science that thrived through the stimulation of the sensorium. In particular, alchemy was reliant upon and had to actively produce a body that was *polymorphous* and *polyphonic*. As such, the corporeal subject was not outside the experiment of alchemical transformation but was part of it. Exemplary in this respect is the German Rosicrucian alchemist Michael Maier's book *Atalanta fugiens* (1618), which was a truly multimedia text weaving together visual emblems, literary analysis of Ovid's legend of Atalanta that revealed allegories for alchemical processes, and accompanying fugues (musical scores for three voices). As Maier writes on the title page of *Atalanta fugiens*, "Accommodated partly to the eyes and understanding, with figures cut in Copper, and Sentences, Epigrams, and notes added, partly to the ears and recreation of the mind, with 50 musical flights of three voices, whereof two may correspond to one single melody Appropriated to be Sung with distichs, not without Singular delectation, to be seen, read, meditated, Understood, distinguished, Sung and Heard."[33] Altogether the book was designed to "teach" the lessons of alchemy through a full-bodied, sensorial engagement with the eyes, ears, and intellect, each revealing nuances of meaning to be puzzled out by studious readers.[34] One had to acquire a taste for the taste of alchemical science. The alchemical ear and alchemical eye had to be produced through practice in order for the intellect to grasp the deeper spiritual and philosophical importance of transmutation.

Citing Maier's work, Alexander Roob describes how alchemical "thought-pictures" were attempts to "reach the intellect via the senses" through a cryptic, pictorial world that demanded forms of sensual thought.[35] Thinking and sensing were not dichotomous (and by extension, competitive) but were mutually innervating and co-constituting. Take for instance Jacobus Sulat's *Mute Book*, which dates from the late seventeenth century.[36] In this book, Sulat (an anagram) created a text of hieroglyphic pictures depicting various alchemical processes on the way to producing the *lapis* or philosopher's stone. The book is mute in that it lacks written text, yet it thinks volumes through the images. Sensual images allegorically reveal intellectual lessons in alchemical arts, while at the same time these lessons ignite the senses through highly evocative imagery. Notice the synesthetic drift of such "reading" practices in which to read is to hear a book speak, while seeing becomes its own kind of thinking. Alchemy is a redistribution of our sensorial apparatus, a refashioning of what the senses can do and how they interact with each other through a creative-descriptive corporeal science. But more than a redistribution, these alchemical texts teach a lesson of calibrating and balancing thinking and feeling, intellectual reflection and sensorial engagement, studious reading/contemplating and empirical experimentation, thus helping the alchemist to avoid extremes that could, in the end, lead to bad abstractions or sensuous indulgences.[37]

Thought was never fully detached from the senses, and instead could take the form of the alchemical vision. Zosimos's ancient text *Visions* offers an obscure recounting of a dream vision full of transformations between body and spirit and death and regeneration. His recipes are stylized in relation to a series of interlocking dream states from which he learns various lessons concerning alchemical procedures. After a horrifying dream of the death of the priest Ion, Zosimos recounts "And being afraid I awoke and thought 'Is this not the situation of the waters?' I believed that I had understood it well, and I fell asleep anew. And I saw the same alter. . . ."[38] Through the dream, Zosimos had various visions that revealed to him secret knowledge. Thus visions/dreams are essential components of alchemical education. Rather than resolve the status of such visions, for us what is important is the way that visions indicate (a) an interest in perceptual modification (repartitioning of what can be sensed, or how sensation can touch that which is insensate), (b) a blurring of distinctions between literary fiction and scientific description, and (c) the educational use of such visions. The body itself was part of the experimental situation, subjected to the same forces as the transmutating metals. Indeed, it is often the case that transformations in metals were expressed in terms of anthropomorphic bodies engaged in "conjunction" or a "chemical wedding" necessary to produce a philosopher's stone, which was then represented as a hermaphroditic offspring. As Roger Bacon cryptically wrote in his alchemical text *Radix Mundi* from the thirteenth century, "The same saith Merlin [the sorcerer and alchemist]: *If you Marry the White Woman* [sulphur] *to the Red Man* [mercury], *they will be Conjoyned and Imbrace one another, and become impregnated.*"[39] Through dreams and visions, bodies and metals enter into strange and uncanny couplings that lack the "objectivity" of modern sciences while at the same time suggesting rigorous investigation, experimentation, and discovery that is often lacking in the "subjective" realm of the arts. Stated differently, alchemy as a corporeal science unites two notions of experience that are somewhat retained in the French word *experience* (yet are divided in the modern, Western world): empirical experimentation and personal experience (as in revelation or visions).

And finally, from the beginning, alchemy was an occult or hidden/secret science that expressed itself through riddles, puns, covert names, and allegories. Famously, alchemical writers often deployed cover names of *Decknamen* to disguise meanings and processes. Quoting Zosimos, Principe notes that the production of transmuting "waters" (as agent of the death of metals and their rebirth through transmutation) is "the manifest secret, that which is studiously hidden."[40] Importantly for us, the secret is *manifest*, meaning it shows itself. In other words, what was hidden in Zosimos's writing points to itself as hidden, so that it can be studied. In this formulation, there is a key distinction in alchemy between secrets meant simply for the alchemist's privileged eyes (and thus face inward, retaining a certain level of cult status or esoteric elitism) and studious secrets that give themselves away for study (and thus face outward toward a larger public). This amounts to a distinction between hidden secrets and exposed secrets. In this sense, we find a tension within the alchemical emphasis on secrets: a tension that expresses itself through intricate alchemical symbols that hide secret knowledge while simultaneously *giving this knowledge away.* This is perhaps best articulated in the medieval text *Libellus de Alchimia* written by Albertus Magnus. At the beginning, Magnus warns that the workers of this art must be "silent and secretive and reveal his secret to no one, knowing full well that if many know, the secret in no way will be kept. . . ." And yet, *Libellus de Alchimia* itself promises to aid its readership to "understand the secrets of this art, and what was hidden for such a long time."[41] The tension between hiding and exposing secrets is precisely what makes such secrets "studious," in need of constant translation, further experimentation, and endless interpretation as it is never certain that the secret has fully revealed itself. As such, Magnus is not simply committing a performative contradiction, instead, his two antithetical injunctions (to hide and to reveal) should be read together in contact with one another as revealing the ambivalent ontology of the studious secret.

There are two important alchemical studious secrets that are worth noting. First, the androgyne or hermaphrodite was a unification of masculine and feminine principles and was often represented as a single figure with the head of a king and a queen. Such figures queer the rise of a stark division between masculine science and feminine art that characterized modern, Western culture by offering impossible solutions (solutions that are not rational syntheses so much as speculative fabulations).[42] Another figure that suspends dichotomies is the paradoxical ouroboros. Timothy Morton refers to this symbol of a serpent eating its own tail as a "strange loop."[43] Drawing on the Norse meaning of "weird" as "twisted," Morton argues that ouroboros twists together concepts that at first blush seem to be opposites, such as life and death, nature and culture, science and aesthetics. Thus, alchemy is a ritual practice for producing strange loops (so many contact points) that make visible the (invisible) monistic substance of reality through transmutation. As Zosimos once wrote, "all things are mingled and all things combine."[44] Alchemical imagery is full of these paradoxical combinations that undercut any notion of a world predicated on strict, fixed boundaries or analytical binaries.

For us, the space-time of the alchemical studio concretizes these underlying features, or brings them to life by providing conditions for spiritual, aesthetic, and physical experimentation. In particular, we highlight four characteristics of the alchemical studio that exemplify its status as an extra-ordinary space of contact between science and art, knowledge and pleasure, head and hand, creating loops that seem to defy the laws of disciplinary gravity that are prevalent today (even in our purportedly interdisciplinary times of hybrid disciplines such as art, science, and technology studies). These features include deconstruction of differentiated spaces (porosity), deactivation of authority (profanation), a-disciplinary experimentation (pliability), and inside-outside/above-below disorientation (contamination). In this sense, we find resources in the alchemical studio for addressing the spatial problematic in current attempts to promote research-creation, and thus resist the reification of space into the artist's white cube vs. the scientist's laboratory. Instead, as we will describe throughout the book, the alchemical studio is a space-time that is dynamic: a precondition of while simultaneously emerging from loops and knots of rules and drift, inoperative works and ritual practices, dark visitations and even darker infrastructures.

Porosity

First, alchemical studios rarely fit nicely into the demarked and separated "functional" spaces of modern institutions or bourgeois households in which science exists in the laboratory and art in the studio (and never the twain shall meet). In the book *Vom hylealischen Chao* (1597), the Paracelsian doctor Heinrich Khunrath of Leipzig chastises those who "utterly un-Philosophically separate *Oratory* [Kabbalah] and *Laboratory* [alchemy] from each other."[45] The workings of the kabbalist and the alchemist are complementary, and it is through one that the other can achieve "analogical harmony" between the macrological/hyperphysical world of spiritual forces and the micrological/physical world of natural experience.[46] In an engraving of Khunrath's *Lab-Oratorium* (1609), the *Auditorium* mediates between Oratory and Laboratory, suggesting that music is the contact point between the metaphysical and the physical, between the verbal activity of the Oratory and the manual labor of the Laboratory (or, in our terminology, between the head and the hand, between science and art, knowledge and pleasure).

The Uffizi Gallery, emblematic of the Florentine Renaissance and a forerunner of the modern museum, also illustrates the porous nature of the alchemical studio.[47] Constructed in the sixteenth century for Cosimo I, studios for art and alchemy were placed side-by-side in the same building complex so as to facilitate collaboration between artists and scientists. A *fonderia* or foundry for the arts using fire was located on the second floor of the Uffizi and was shared by goldsmiths, alchemists, metal workers, and chemists, all of whom were engaged in different labors of "distillation" that utilized overlapping materials and equipment. Galleries, forges, workshops, and collections comingled so that the system of correspondences defining the natural world could be mapped. Although Ferdinando I later transformed the *fonderia* of the Uffizi into a court workshop for economic and diplomatic purposes according to principles of rationalization and optimization, at its foundation, it was explicitly concerned with experimentation and research that gathered together *naturalia* and *artificialia* into a spatial-temporal encyclopedia.

Alchemical studios offered repartitionings of space, creating their own, ambiguously situated spacings that were simultaneously a part of and apart from the household.[48] At the same time, the spatiality of the alchemical studio, in particular, had a tendency to interrupt or suspend the functions of traditional rooms, popping up in bedrooms, kitchens, and so forth. While there are many historical examples of this phenomenon, surrealist painter Leonora Carrington—whose practice was greatly influenced by alchemical symbolism and was referred to as the *femme sorcière* by other surrealists such as André

Breton—offers a modern exemplar of the peculiar spatiality of the studio.⁴⁹ In 1943, she moved to Mexico where she became involved in an existing émigré surrealist group, including Remedios Varo (whom we will discuss in the vestigial comments at the end of this book).⁵⁰ Together, they began an intense artistic relationship, which involved pseudoscientific inflected investigations/rituals into cooking, transforming Varo's kitchen into an alchemical studio. In the studio, they were interested in establishing connections between women's traditional roles in the household and alchemical acts of self-transformation. This practice unlocked new potentialities within otherwise oppressively gendered spaces for radical reinvention.⁵¹ Around the same time, the photographer Edward James described Carrington's art studio as follows: "The place was combined kitchen, nursey, bedroom, kennel, and junk store. The disorder was apocalyptic: the appurtenances of the poorest. My hopes and expectations began to swell."⁵² The alchemist's studio is a place of disorder (it does not organize itself according to common sense logics) that redistributes who can do what when and how. Suddenly distinct spaces implode, disorganizing the strict choreography between bodies, gestures, equipment, and spaces that is often ascribed to specific rooms of a house. While this might be "apocalyptic" or anarchic, James rightly finds a new intensity in the mishmash, which swells out of him with a certain amount of unexpected exuberance. Similarly, Carrington's visual map of her book *Down Below*⁵³—which chronicles her time spent in a mental asylum after the arrest and internment of her lover, Max Ernst—morphs the oppressive space of the asylum into an alchemical studio coded in various symbols such as crescent moons, gates, caves, spirals, snakes, and orchards all of which disorder the disciplinary space of the asylum as well as its economy of surveillance. Indeed, Carrington's mapping reveals the politics of occultism as an occlusion in the eye of the surveillance apparatus, reminding us that the invisible and opaque is often a tactic of survival for the oppressed. The map also deconstructs boundaries between internal, psychic space and external, physical space (providing a spatialization of a corporeal science that is nomadic, deterritorializing the striated space of the sanitarium as a line of flight). It is a schizoanalysis of the space of the asylum as crisscrossed with gates and apertures that do not abide by the binary logics of inside (psychological phenomena) and outside (objective reality). The public and private, inside and outside, self and other, psychic and physical are demonstrated to be porous, if not nebulous and arbitrary, demarcations that are easily subverted through alchemical drift.⁵⁴

Profanation

In certain cases, the alchemical studio could deactivate relationships marked by authority often exemplified in terms of the dominance of the *paterfamilias* in the household and his household duties. The alchemical studio, in this sense, was not only a space for experimenting on and with materials and ideas but also with familial and social relations. For Gadi Algazi, many historic paintings of the alchemical studio attempted to depict alchemists as "bad scholars" and the paintings were meant to be read as moral lessons on how *not* to run a household. Children, apprentices, servants, and wives seem to mix indiscriminately in spaces that lack clear partitioning that would separate the studio from the kitchen and the work of child-rearing from the experimentation in the scholarly study. As Algazi writes, alchemical studios were often depicted as "crowded, shared, and often untidy spaces" that had the peculiar effect of "confounding cultural categories and mixing separate domains in domestic settings,"[55] making possible a reconfiguration of roles between husbands and wives, adults and children. Thus, we can read these images against their intended meanings and see such anarchic disruptions as indicating a potentially liberatory freeing up of social relations from prescribed categories, roles, and hierarchical divisions of labor. Our blatantly ahistorical and heretical interpretation of these allegorical images finds within them the potentiality for queering the family unit, undoing the lines that direct certain bodies in certain directions toward certain objects. Instead, bodies, practices, and objects generate queer contact points that would have been offensive to the ethical sensibilities of reasonable, enlightened, bourgeois audiences. Again, we are reminded of Carrington's studio, which hacked into various gendered (the kitchen) and oppressive (the asylum) spaces in order to release alchemical potentials where they were not supposed to be, and indeed, might even signal a breakdown in the control over space that is essential to any patriarchal and/or disciplinary regime of power.

Perhaps more interesting still are the ways in which women's work and children's play were integrated into the verbal and visual languages of alchemy. In many images, women's labor such as washing the linens, cooking fish, and tending to a fire are all taken as paradigms for alchemical labors. In this way, divisions between women's and men's work are complicated if not overturned. Male alchemists ought to apprentice themselves to women in order to understand their own arts. Images of children at play in or around the alchemist's studio also became pedagogical models for an advanced stage of experimentation in which the alchemist's labors were complete, leaving free time (play). In both cases, alchemy *aligned itself* with domestic labor (women's work) and play (children's activities), effectively queering the figure of the alchemist who no longer operated according to definitions of masculinity.[56] In the book titled *Gloria Mundi*, written in 1526, the philosopher's stone is described as "familiar to all men, both young and old; it is found in the country, in the village and in the town, in all things created by God; yet it is despised by all. Rich and poor handle it every day. It is cast into the street by servant maids. Children play with it."[57] In other words, the "secrete" of alchemy is not the special knowledge of an elite group of men so much as *common knowledge* that is found in the average, everyday lives of men, women, and children.

In short, the alchemy studio contained multitudes. As opposed to humanist images of the lone, male scholar, detached from worldly affairs, here thinking and making converge through social discourse and intercourse or what can be called "convivial" studies or studios.[58] The alchemical studio, in particular, was an assemblage of bodies (male and female, adult and child), speeds (the hyper-activity of playful children, the slow concentration of men and women in various forms of domestic labor), and affects (which seem to circulate between bodies and drift outside the physical space of the studio into the surrounding environs). Strangely, this same point would be reiterated centuries later by Deleuze, this time in relation to Godard and the film studio. "When Godard says he would like to be a production studio, he is obviously not trying to say that he wants to produce his own films, or he wants to edit his own books. He is trying to say just ideas, because, when it comes down to it, you are all alone, and yet you are like a conspiracy of criminals. You are no longer an author, you are a production studio, you have never been more populated."[59] When making a film, Godard ceased to be an individual author and instead was a becoming-studio. And this becoming-studio is a criminal act, a fugitive act, not unlike the anarchic alchemists of yesterday who ought to be managing the household or taking care of official business rather than experimenting with the all-too-feminine and all-too-childish experiments in transmutation.

Pliability

Like the books of secrets, alchemical studios were spaces that were decisively a-disciplinary, meaning that they did not clearly separate out science from art, knowledge from pleasure, hand and head, and instead were fertile contact zones. In the book *Alchemy: Its Science and Romance*, J. E. Mercer points out that the typical medieval alchemical studio "was a place for work. That is certain. But it was much more than this. It had an air of mystery; its furniture was decked with signs and symbols that appealed to the imagination and the emotions . . . that is to say, the labour was not undertaken in the hard clear light of science, but in an atmosphere of mystery and devotional fervour"[60] that defines the hermetic arts. Mercer emphasizes the cultivation of a particular atmosphere that is essential to the experiment itself. Such atmosphere might be activated through various ritual practices including chanting incantations, as handed down from Egyptian and Greek alchemists. Occult atmospherics were not merely decorations tacked on to more essential scientific methods of investigation. They were *essential* to the success of the experiment itself, and thus part of the genuine "labour" of alchemy in the medieval and early modern periods.

Take for instance the writings of George Starkey, often considered the "last" alchemist right on the cusp of the modern era, and his detailed descriptions of experiments in his studio. Historically, notebooks written by alchemists have been largely dismissed as nothing more than occultist flights of fancy, yet, as William R. Newman and Lawrence M. Principe demonstrate, Starkey's esoteric symbolism, emphasis on spiritualism, insistence on "visions" as key to experimentation, and opaque narrative belie a keen interest in testing and experimentation often more associated with chemistry than with alchemy.[61] His work sits in a liminal zone of contact opened up between the authority of the text, the authority of the experiment and of observation, and the authority of the divine, which he sometimes claimed access to. Scholastic interpretation, poetic invention of riddles and allegories, divinatory philosophy, and scientific experimentation bled into one another almost indiscriminately, thus disturbing modern tendencies toward separation and specialization.

Contamination

And finally, the alchemical studio undermined not only the internal structure of the home but also the separations of the private and the public and nature and culture. While alchemical studios are often thought of as hermetically sealed, esoteric, secret spaces, Christopher Wood argues that studios, in general, are "nonplaces" in constant dialogue with the pasture (the wilderness, the outside), the medieval workshop, the pharmacy, and the modern university and thus could not fully separate themselves from the social and natural commons.[62] The notary Lapo Mazzei (1250–1412) from Prato compared his feeling of being in his studio to being "as happy as the good hermits are on a mountain."[63] Even in their attempts at secrecy and hiddenness, the alchemical studio inscribes its outside within its inside, demonstrating its aporetic spatial structure. Interestingly, in John Amos Comenius's description of work in the studio from his illustrated children's textbook *Orbis Sensualium Pictus* (1659), he highlighted the work of the scholar as marking "beast things" in a book with a "little Star, in the Margin."[64] Thus in this humble and largely ignored gesture of the scholar, we find a practice in which the outside is brought inside the studio while also bringing what is above down to earth, thus fulfilling the alchemical maxim purportedly written by the mythical figure Hermes Trismegistus: "that which is below is like that which is above."[65]

These four dimensions of the alchemical studio all concern porosity, profanation, pliability, and contamination as essential features of a space a-part of other spaces. The studio contaminates inside and outside, nature and culture; it profanes the spaces of the household (the bedroom and the kitchen) undermining the gendered division of labor which they prescribe. Another way of stating this is that the alchemical studio is never clearly defined by strict boundaries, and, in a sense, is constantly inducing drift and is a product of drift. Hence our insistence that the space of the studio is in essence also and equally a *movement* of becoming-studio or studioing.

The alchemical studio as a space and time for movements of drift stands in stark contrast to the rise of the modern laboratory as an anti-drift spatial formation concerned with control. In 1597, Andreas Libavius published *Alchemia*, a book dedicated to systematically teaching "chymistry" without the spiritual dimensions of alchemy or the influences of Paracelsus (one of the ancient sources of alchemical knowledge).[66] In the second edition of the book, Libavius included a schematic design for a "chemical house" or blueprint for constructing a building to house chemistry laboratories. This blueprint is considered one of the first attempts to define the modern, European laboratory and thus is an important point of comparison with the alchemical studio. To begin, the chemistry laboratory was, for Libavius, to be separated from the rest of the household. It demanded its own house. The history of the laboratory thus concerns separation, specialization, compartmentalizations, and clearly demarcated boundaries (rather than the somewhat anarchic drift found in and around the alchemical studio). Opposed to the secrecy of the alchemy studio, the chemistry house was to stand at the center of town and would have been open for all to see. Thus the esoteric, secretive, and occult nature of alchemical studios is absent. There are not secrets to study (and give away) in the laboratory. Instead, Enlightenment transparency and reason are embodied in an architectural form. While the chemist did have a private laboratory linked to his living quarters, it was less a secret chamber than a precursor to the university lecturer's laboratory in the nineteenth century. Also of note is how each room in the house served a specific function within the overall division of chemical labor. There were rooms for specific operations such as the analytical laboratory housing assay furnaces, the coagulatorium with tubs and vats, a preparation room, and a pharmacy for making medicine. The laboratory house was rationally partitioned, each room defined by a specific function within an overall division of labor in the name of increased precision, exactitude, control, and measurement. While these elements might have been present in the alchemical studio, Libavius's plan illustrates a qualitative shift toward an increasingly determinant break with what was perceived to be alchemy's dubious past but also with the uncomfortable comingling of knowledge and pleasure, science and art, reason and taste. And finally, the plan as laid out by Libavius is an abstract plan, separate from the specificities of an individual chemist or concrete context. It is a disembodied blueprint, detached from the idiosyncratic ritualities of alchemists who deal with allegories, riddles, and recipes that reside in the dark rather than in the light of reason. In short, the laboratory becomes increasingly rationalized, functionally subdivided, abstractly disembodied, and thus detached from any specific place or time. The a-disciplinary corporal science of alchemy and the boundary defying particularities of its atmospheric studio space-time become coded as archaic embarrassments. What remained of the studio was relegated to the arts and the alchemy of creative genius.

The following book offers a spatial-temporal *retrieval* of the vestigial potentiality of the alchemical studio as providing the hospitable conditions for experimenting with research-creation that does not fall into the trap of (a) merely submitting the arts to the reign of the sciences or (b) hybridization. This is a dark space-time that drifts in the zone of contact between art and science, studio and laboratory (and auditorium), head and hand, pleasure and knowledge. Claiming the alchemy of research-creation within the space-time of the studio enables us to safeguard the possibility of *creation* in research-creation as a study practice capable of fundamentally rupturing the binaries that structure the modern, capitalist world and its sciences.

Entering the Alchemical Studio

As we will argue, the alchemical studio suspends the division between art and science that supports the division between knowledge and pleasure, head and hand. It is not a zone of hybridization between two existing terms placed in a relation with one another but rather a line of contact that suspends what we think of as science and art to begin with. As such, we are not providing yet another (neoliberal) call for more interdisciplinary research, which we read as maintaining the disciplines and picking and choosing elements that are mutually beneficial, productive, or profitable (for capitalist value creation). When entering the alchemical studio, the very coordinates orienting us toward and around certain knowledges and their practices at the expense of others are neutralized, creating a queer set of (dis)orientations that will not straighten out, as Sara Ahmed would say,[67] but prefer to drift sideways and to twist into strange loops. *Ouroborotics* (as a combination of the alchemical serpent ouroboros and robotics/erotics) is another name for alchemical studioing practices that study how potentiality drifts from its designated, predetermined destinations in order to twist away from existing functions within the order of things. Stated differently, alchemical studioing is a space-time for producing works that *do not work* (are inoperative by common sense standards) but for that very reason, embody a potentiality that has yet to be claimed by either art or science (or some combination of the two): a potentiality that remains in potential, that remains without a destination, that has no home, that has no reference (yet), that is decisively queer.

A studio, or better yet the practice of space-time warping through studioing, makes room for perceptual deviancies, disjunctive senses, and drifts of thinking. Studioing thus produces fabulations rather than verifiable knowledge claims. For Stengers, fabulation is not a break with reality so much as it is a way to make something differently perceptible[68]—to make visible the invisible and oblique dimensions of reality, or the accursed share that falls outside of the scope of our present perceptual capacities, or the inoperative waste/by-products of productivist paradigms. Or as Agamben might argue, what is at stake in a studious fabulation is a pleasure that knows and a knowledge that enjoys, which would necessitate (a) a repartitioning of senses through (b) practices that are not reducible to science or art that (c) experiment on inoperative objects, gestures, and ideas within the emergent space-time of the studio for (d) the transversal contact between sensations and concepts, which alchemically swerve into one another, collide, and create singularities out of darkness. The movement of studioing is essential for avoiding various traps within research-creation (as outlined above). On our view, the drift of studioing sets art and science adrift from their disciplinary boundaries and hierarchies so that they come in contact as strangers, not simply to each other *but also to themselves*, alchemically altering their composition (what they can do, say, or how they can appear). The result of such research-creation is precisely *new flavors* of knowing how to feel and feeling one's knowing.

What kinds of queer (dare we say, antifascist, anticapitalist, and neurologically diverse) flavors are produced in the space-time of the alchemist studio as both place and movement within/constituting such a place? This question drives our project toward uncertain and open-ended exploration.

The present book outlines the various alchemical features of the studio as a profane space-time for research-creation. To do so, we draw upon an eclectic group of artists, philosophers, and esoteric theorists (not that these categories will make sense for much longer) that are all, directly or indirectly, influenced by alchemical practices and literatures, or animated by more vague and intuitive impulses linking them back to ancient, esoteric precedents. Some directly weave alchemical and pataphysical references into their practices while others do so only indirectly (perhaps invisibly if not for the dim perceptual fabulations offered by this book). We include various visual artists and poets, surrealists and pataphysicians alike: those who engage with the occult potentialities of alchemy through a suspension of good and common sense. Some engage with the unconscious (surrealists), while others push reason and rational calculation to its upper limits (sur-rationalists or pataphysicians). We liberally include all in the copious space-time of studioing, or rather, we attempt to recognize in their work, alchemical dimensions that escape the science/art, studio/laboratory divides. What follows is a profane concoction that does not abide by strict academic norms and exchanges verifiable influences for chance encounters and happy accidents. But this eclecticism enables us to untether theorists and artists from ideological or epistemological camps binding them to this or that school of thought or this or that entrenched set of problematics. The resulting drift enables fabulations to emerge that are cryptological, meaning experimental and imaginative, pointing toward queer, dark potentialities that might otherwise remain concealed within the prison house of disciplinary specializations. Such drift does not erase differences, nor does it indulge in sloppy connections. Instead, it investigates a certain force that pulls an experiment in thinking, feeling, and sensing in directions that cannot be easily mapped onto accepted ways of framing what counts as scholarly rigor or accuracy in advance of its own happening. Thus the "logic" of any given constellation of ideas and theories presented in this book must be thought of as singular happenings that ought to be approached contextually (or immanently) in order to gauge various swerve effects on knowledge and pleasure.

In sum, returning to the latent potentiality within the alchemical studio enables us to ask a question that we feel is essential yet undertheorized in terms of research-creation: How can research-creation avoid classical conceptions of space that are too static but also postmodern, accelerationist notions of space that are too dissolved? The alchemical studio forces us to focus on this question. The studio is always a practice of studioing, or of movement that embodies both the line of drift and the clumping mass of such drift as things accumulate or coagulate. Studioing is a shifting, permeable configuration of protocols, rituals, recipes, and works that continually produce their own space-time in which dichotomies are neutralized, opening up the potential for new flavors to flourish in what would otherwise be forgotten dark corners. The resulting notion of research-creation that emerges from within the alchemical studio will challenge the ways in which research-creation has thus far been conceptualized (and allied with the sciences).

To engage in research-creation as a form of alchemical studioing is to produce an alternative geography of bodies, disciplines, and landscapes that is cryptic and dark. The *English Heretic Collection* by Andy Sharp is a case in point. Sharp channels various esoteric knowledge systems, pataphysical humor, situationist psychogeography, and alchemical references into a project that undermines the consumerist history of English heritage by setting up Black Plaques honoring obscure and mostly forgotten occult figures. But what is important here is how Sharp refers to this process as one in which the countryside is transformed into "studio" space, or an "arena for disrupting reality," that provides space-time for "an open-ended and ludic ritual in all dimensions."[69] In other words, the unique space-time of the studio provides a suspension of the operative logic of heritage geography through the production of occult gates marked by sonic signatures, weird tales, wyrd sciences, and Black Plaques—a witch's brew in which knowledge and pleasure, art and science, head and hand enter into queer relations of contact destabilizing the coordinates of tamed, commodified "heritage" sites.

This book is itself a kind of psychogeographical landscape, with each chapter actings as an alchemical gate. Yet, these are not gates to a cathedral, temple, or palace—all of which Georges Bataille argues are manifestations of a collective desire for authority and social stability.[70] Rather, these are gates to a ruinous crypt, a partially raided tomb, an abandoned mound. As one travels through these gates, one does not enter into increasingly secret rooms that culminate in an inner sanctum containing mystical treasures. While this might be the typical image conjured up by reference to esoteric arts, it is an image that ought to be overturned for its elitism and potential for abuse. Instead of traveling inward to approach higher-order truths reserved for the exclusive inner circle of initiates, the gates of this book are porous (leaky, perforated, ruinous, letting the content overflow and escape any firm container), profane (the "truths" revealed are never sacred, the grounds never holy), contaminated (promoting contact amongst strangers rather than separation), and pliable (moveable apertures, holes, escape hatches). Thus, the structure of the book is one in which gates are entered only in order to discover that one has already exited, that each interior room encountered is already its own outside, that each door closed is already a threshold to another opening. The ending of such an experiment cannot be anything other than a loosening of the gates at the precise moment when one might feel that some "progress" has been made or some "conclusions" have been reached or some wisdom has been "discovered." These are, in short, gates that are inoperative—they neither keep out nor keep in . . . they merely mark a threshold or zone of potential contact with dark precursors. The alchemy of research-creation cannot happen in cathedrals, temples, or palaces but rather in places where the architecture of authority crumbles and sinks into the crypt, leaving behind idle remnants that can be picked up and used for *God knows what*.

Before one takes the first, tentative steps into this ruin called a book, we ask the reader to pause in our "antechamber" in order to consider several criticisms of alchemy (in particular) and the occult (in general) as well as our own (self-created) rules for entering into the darkness of these practices and discourses. This book is most certainly a defense of the alchemical dimension of research-creation, but it is far from a simple and naïve celebration of the occult and the esoteric. As such, we must pause before entering the first gate to consider what rules might be helpful for entrance. These rules are just as much for the reader as they are for us, the authors. They are rules that concern both how to enter the gates, but also how we have (de)constructed the gates.

Then, gate one opens upon the crypt, revealing the studio as dark infrastructure, or an infrastructure that is (a) secret but also (b) useful for conjuring dark precursors (clinamina or swerving atomic ghosts in the ether). The gate includes two examples of dark infrastructure: the fireplace and the window. Through detailed analyses of Baccio Bandinelli's "Night Class" (mid-sixteenth century) juxtaposed next to Betye Saar's "The View from the Sorcerer's Window" (1966), we argue that dark infrastructure is not so much an amalgam of specialized instruments of the alchemist but rather can be anything at all (given specific conditions and constellations of things, forces, gestures, and movements). What matters is therefore *when*, *where*, and *how* infrastructure is taken up and put into practice. Gate two focuses more closely on the composition of the studio as a confluence of rules (immaterial dark infrastructures) and drift that form loops and ultimately knots (impractical ritualities of studioing) that are necessary for creating hospitable conditions for dark precursors. Through a somewhat labyrinthian structure, we attempt to unfold the recursive movement that leads from rules to loops, loops to knots, knots to more rules, and so on. We then pivot to an example from our own university teaching practice in which the classroom is transformed into an alchemical studioing space-time machine through the adoption and implementation of a loose set of rules that set in motion physical and conceptual drift amongst students. The third gate turns our attention to the effects of studioing on the alchemical practitioner who must take certain risks. These are not so much risks of failure (which would presume the possibility for accounting/measuring progress or regress) as they are risks at being perceived of as awkward, weird (twisted), "neurodivergent," or "mad" for taking up drift. To demonstrate, we turn to several practices of research-creation, including the contemporary performances of Pilvi Takala who disrupts the office space by preferring not to work, the social choreography of Michael Kliën, and Leonora Carrington's memoir of time spent in a mental institute. Before exiting the gate, we argue that such risks are unique to the studioing practices of research-creation as opposed to both the creative economies of capitalism (which, in the end, concern the regulation of risk) and the destructive creativity of fascism (which risks everything, ultimately resulting in suicide). The fourth gate turns to the question of the politics of studioing and how drift is essentially an antifascist space-time concept. In particular, we offer a synthetic looping together of various theories of fascism inspired by Walter Benjamin, Georges Bataille, Gilles Deleuze and Felix Guattari, and finally Fred Moten, whose work grounds insights offered by the aforementioned authors specifically within what we will come to call the antifascist pataphysical alchemy of blackness. This concept is then brought to life through our reading of Sun Ra's film *Space Is the Place* (released in 1974), which does not simply illustrate a theory so much as pole-vaults it to another level of cosmically sonic experimentation. The final gate positions the problematic of the studio in relation to postdigital research-creation, which in turn allows us to once again return to the body of the alchemical pataphysician and the ways in which drift mutates the body through the production of vestigial organs. As both exemplar and as warning of the potentially catalyzing and manipulative/exploitative effects of postdigital research-creation, we offer a detailed engagement with David Cronenberg's student film *Crimes of the Future* (1970).

In sum, the gates of this book offer a loose architectural frame, or dark infrastructure that embodies a certain level of *organizational drift*. Each gate offers a strange mapping of juxtaposed fragments, profanations of academic forms, and looping digressions that are accumulations (clumpings) of drifts that have temporarily stabilized into the porous, profane, pliable, and contaminated space of our imaginary studio (a house of cards, propped up precariously through certain dark forces that we ourselves do not always understand). And we hope that like all studios, this one will provide an opportunity to set adrift what can be thought and what can be sensed.

Antechamber

The Alchemical Occult:
A Warning and a Promise for Experimental Life

While alchemical studioing might offer a unique resource for rethinking the spatial and temporal dimensions of the divisions between pleasure and knowledge, art and science, head and hand, the relationship between alchemical practices in the studio and the occult might give some readers pause. Certainly there is a long literary and scientific tradition of critiquing alchemy that begins with Chaucer's "Canon's Yeoman's Tale"[1] and then extends into the eighteenth-century critiques of alchemy by chemists who complained that alchemists were nothing more than charlatans peddling various "get rich quick" tricks.[2] Until the beginning of the Enlightenment, theologians and ecclesiastical historians had various opinions on the relationship between religion, philosophy, and paganism. In other words, they had to take the presence of esoteric knowledge traditions seriously, either advocating for some kind of synthesis between Christianity and paganism or arguing against paganism in the name of a pure Christianity. But starting in the eighteenth century, this changed dramatically with the arrival of two new paradigms. From the "religionist" point of view, esoteric knowledge was simply irrelevant to the study of Christianity and thus could be dismissed without further comment. And from the Enlightenment paradigm, reason was employed to measure the stupidity and foolishness of esoteric doctrines and occult practices inherited from a pagan past. In the name of purifying philosophy of the taint of irrationality by building firm walls between art and science, pleasure and knowledge, head and hand, Enlightenment authors provided the essential intellectual architecture that would act as the foundation for the divided world. The second paradigm in particular became increasingly dominant, and, in many ways, remains active today, and is a key reason why the esoteric and occult remain largely tabooed topics in the academy.[3] We can hear the echoes of the eighteenth-century dismissal of pagan esotericism when members of the enlightened class of academic intellectuals say, "Who would want to doubt the rational critique of the alchemy studio and the embrace of the virtues of reason, transparency, and public engagement promised by Libavius's chemistry house?" Good, common sense would dictate that alchemy ought to be relegated to the dustbin of history as a relic of a superstitious, and thus dangerously delusional, past.

There were also sociopolitical reasons for being suspicious about the occult. Marx famously wrote that the bourgeoisie had become "the sorcerer, who is no longer able to control the powers of the nether world whom he has called up by his spells."[4] The liberation of productivity from the grips of feudalism's constraints and the emergence of a rational division of labor turned into a Faustian bargain with a seemingly unstoppable and increasingly monstrous economic system with endless expansive (and exploitative) powers beyond the command of all the liberal (bourgeois) economists trying so desperately to control its momentum and velocity. The theme of necromancy also informs many aspects of Marx's detailed analysis of the structure of the commodity, commodity fetishism, and the metamorphoses of money that perpetually shift between money as commodity and money as a representative of exchange-value. Marx summarizes the various metamorphoses of capitalist exchange as follows: "everything becomes saleable and purchasable. Circulation becomes the great social retort into which everything is thrown, to come out again as the money crystal. Nothing is immune from this alchemy. . . ."[5] Capitalism is a science of transmutation that unlocks a metaphysical, abstract power within objects known as exchange-value. In this sense, political economy is just a branch of other occult arts. Given this critique of capitalism, it might at first seem absurd to hypothesize that Marx's notion of the utopian-socialist body composed of a willing hand, a ready mind, and a joyous heart could be incubated in the space-time of the alchemical studio. Indeed, it would seem that with the dismissal of the esoteric as pre-Enlightenment stupidity, the esoteric did not disappear but rather took on a new name: political economy!

Drawing inspiration from Marx, critical theorists like Theodor Adorno worried about the rise of occultism through a blend of ancient rites and mythologies wedded to capitalist consumer culture and advertising. For Adorno, the resurgence of the occult in postwar American popular culture was a "symptom of regression in consciousness,"[6] which, if left unchecked by critical self-reflection, could transform into yet another manifestation of fascism. The cause of this symptom is structural (rather than merely psychological) as it reflects the "fetish character of commodities: menacingly objectified labor assails him [the worker] on all sides from demonically grimacing objects."[7] The worker forgets that objects are products of labor-power, and instead they take on a fantastical life of their own, propelled by a semiautonomous vitalism. The world as such becomes animated with spirits rather than with historically concrete contradictions. As thought becomes magical, it is, for Adorno, "assimilated to late capitalist forms" losing its critical capacity to resist and speculate beyond the present.[8]

Thus, Adorno would not be surprised when the contemporary occult guru Richard Metzger begins his edited collection of essays under the pseudo-radical title *Book of Lies: The Disinformation Guide to Magick and the Occult* with the following: "Some sorcerers use painting or music of fiction to work their magick, but I quite like the idea of having a 'magick business' — both *literally* and figuratively — as the canvas that I perform *my* magick on. It works on a lot of levels, metaphorically speaking, for me to consider myself to be a *magical businessman*, if you see what I am saying."[9] In the same collection, Grant Morrison highlights the connections between sigils and logos, calling for late capitalist magicians to invent their own brands and "see how they interact with other corporate entities" such as the Coca Cola "spirit."[10] Capitalism and Metzger's magik or Morrison's corporatized sigil experiments are cozy bedfellows precisely because capitalist production and consumption are, as Marx argued and Adorno reiterated, fundamentally propelled by practices that take on magical properties.

While the esoteric has now become fully integrated into the global capitalist market, the very same esoteric traditions are being taken up by fascist movements as ways to resist this form of economic integration.[11] Jacob Christiansen Senholt highlights the ongoing interrelationship between far right, neofascist movements in Europe and esoteric belief systems and occult practices.[12] This interrelationship takes many forms, ranging from rather superficial adoption of esoteric symbols and imagery to the more ideologically complex interweaving of nationalism with the folk desire to return to prehistoric/archaic Nordic civilization. Furthermore, many such movements advocate a distinctly pagan and mythical mentality outside of secular modernity, Enlightenment reason, Christian morality, and multicultural policies. Such fascist countercultures utilize esoteric imagery and beliefs to help create a grand polemic narrative against Western modernity, complete with an alternative cultural genealogy and list of martyrs who have died in the name of preserving a purified Europe.

The strange overlap between the commodification of the esoteric within capitalism and far-right occultism against liberal, mainstream culture can be found in the case of the former academic Nick Land and his work with the Cybernetic Culture Research Unit (CCRU) at Warwick University in the late 1990s. One might argue that the CCRU's experimentation with electronic dance music (especially the techno-rave collective Orphan Drift), cybernetic science fiction, Deleuzian philosophy, and occultism could be viewed as a paradigm for the kinds of research-creation we are calling for in this book. But once it had lost its academic affiliations, the CCRU spiraled downward from a leftist, democratically open group to a rather nihilistic, post-leftist organization intent on spreading conspiratorial myths and superstitions, and Land himself ended up inspiring a new generation of online fascist youth mesmerized by his intoxicating, esoteric, drug-induced hallucinatory jargon, the aesthetics of a Dark Enlightenment, and the speed of techno-capitalist production — all of which combine to accelerate humanity toward a cataclysmic yet liberatory end.[13] In a shockingly fascist turn, Land has asserted that one of the big lies of the globalist liberal elite is equality amongst races. Opposed to any claims to equality, Land has argued that race is a biological fact (not a social construction), that intelligence is not equally distributed across races, and that the more leftist ideology asserts equality, the faster society will collapse. Thus, white supremacist backlash against the left is a key accelerant according to Land, which ought to be used strategically (as the ends seem to justify the means) by the far right. Such logic has been quickly picked up by fascist, online groups, which now view Land as a prophet of an impending race war.[14] Here, Land falls back on modern fascist narratives of race as a genealogical fact and of races as distinct, hierarchically ranked groups. And in other contexts, Land has praised the eugenic tendencies of space colonization in which (predominantly) white, technocratic nerds are given unique privileges for imagining a future that excludes the masses, which Land describes using highly racialized, subhuman, and derogatory terminology.[15] To his credit, Land clearly sees the internal esotericism of capitalism's logic and the external esotericism of neofascism as interconnected phenomena that can catalyze one another toward the destruction of civilization, but for him, this confluence is a solution rather than a problem. In such instances, Land recalls any number of rash intellectuals in the twentieth century who were persuaded by fascism's alluring mix of vitalist, cosmic energies, occult mysteries, and racial mysticism.[16]

There is much to be praised in the ardent defense of the powers of critical self-reflection to resist temptations toward magical thinking as a reflection of capitalism's own necromancy of commodity production and consumption. Indeed, the concerns espoused by Marxists and critical theorists are now more prescient than ever given the rise of global fascism in the twenty-first century from within neoliberal capitalism. In such a situation, it seems particularly dangerous to call for occult experimentation.

Yet we would also argue that any critical polemic against the occult could very well topple over into a positivistic fetishization of "just the facts," which, in turn, would replay the paradox of Western modernity's desire to create a divided world as outlined in the initiation to this book. Indeed, there is a danger in strict dichotomies between facts and fictions, even if such dangers might be strategically outweighed by an overwhelming desire for a sense of security and/ or objective grounds for asserting the fact (and just the facts). Specifically, there might be another way to engage differently with alchemy that does not simply dismiss or dogmatically endorse its occult dimensions.[17] It is important to note that the criticisms of alchemy notwithstanding, alchemy has been an essential inspiration for critical theory (even if it is an often shamefully disregarded one). In this sense, it is not the occult as such that is to blame for the mystifying language of the commodity and its political articulation, fascism, but rather *how* we engage with such hermetic concepts, practices, and spatial-temporal formations. For instance, the surrealists' interest in alchemy was coupled with a deepening study of Hegel, a growing appreciation for Marxism, and political commitments against colonialism. This conjuncture is, on our reading, not a coincidence but rather an example of how occult interests cannot be strictly identified with capitalist economic abstractions or its fascist politics.[18] To further demonstrate this point, we can briefly turn to three anticapitalist and antifascist thinkers and their engagement with (rather than opposition to) occult influences — three thinkers that are key influences on this book project.

Despite his criticisms of spiritualists such as Rudolf Steiner and the ways in which they transform "the commodity into an arcanum" and sell "the arcanum as a commodity,"[19] Walter Benjamin engaged in a productive (rather than merely negative) dialogue with esoterica and occult forces in order to find techniques and forms for articulating a critical theory that could act as a bulwark against fascist tendencies.[20] Starkly opposed to Adorno, Benjamin went so far as to write a "prolegomenon" to a "rational astrology" that would not sacrifice what he describes as the "mimetic genius" of humankind (more on this below) found in pagan culture, children's play, and pseudo-sciences such as graphology (an interest he shared with Steiner).[21] And this redemption of the occult was not a one-off for Benjamin. Instead, the occult bookends Benjamin's career in interesting ways. In his first major research project on the German *Trauerspiel*, Benjamin draws inspiration from premodern astrology. He writes, "Ideas are to things as constellations to stars."[22] Ideas, which are invisible and intangible, only appear through a constellation or "virtual arrangement of phenomena."[23] Ideas are not given in reality (the world of phenomena) but emerge as an image through the gathering together of seemingly far-flung and unrelated elements (much like stars in different galaxies strung together into a constellation). Concerning alchemy, Principe writes, "premoderns tended to conceive of and visualize the world in multivalent terms, where each individual thing was connected to many others by webs of analogy and metaphor,"[24] which stands in stark contrast with modern conventions that compartmentalize, fragment, and isolate the world in terms of disciplines (each of which has its own notion of truth, modes of inquiry, and preferred objects/phenomena). A key tenant underlying hermetic traditions concerns the ways in which individuated things/phenomena/knowledges gain their meaning and significance in relation to a whole and, vice versa, how the whole is contracted within each individuated thing. The same can be said of Benjamin's constellational thinking in which the idea of the *Trauerspiel* is built up of many disparate elements that hang together around an invisible center (a truth) that cannot be grasped directly by the conscious mind but only indirectly through the harmonious concord between various citations that all point beyond themselves toward a hidden (esoteric) horizon. In Benjamin's "monadological" thought,[25] the totality is invisible yet latent within each particular, which must be carefully assembled into a constellation so that the truth may be illuminated in a flash of recognition. Likewise, Benjamin's intention to write a book that contained nothing but citations[26] has ancient roots in the alchemical *florilegium*, which literally means "gathering of flowers." These books were compendiums of quotations taken from many authors concerning alchemical theory, recipes, and cryptic sentences requiring interpretation. While this might seem like nothing more than a confusing conglomeration of randomly assorted bits and pieces that ought to be kept separate, the alchemical *florilegium* had its own aesthetic and hermetic logic that proceeded by way of analogy and intuition rather than disciplinary division. For us, Benjamin's study of the Paris arcades is a similar enterprise, resulting in a sprawling collection of loosely organized themes, quotes, thoughts, and theses.

In all cases, Benjamin does not simply dismiss the occult but rather engages deeply with these traditions as models to inform his own critical practice and his experimental forms of writing. Whereas Adorno stepped outside the alchemical studio, Benjamin entered into it to retrieve inspiration for cultivating an unnamed science. Of course, there were substantial worries by his Frankfurt School colleagues that Benjamin, at times, went too deeply into the occult, especially in his engagement with protofascists such as Ludwig Klages and Carl Jung, but for Benjamin, the powers of the occult could not be overcome through negation and instead had to be reconstructed (reconstellated) through redemptive critique.[27] For instance, Benjamin described Jungian psychoanalysis as a "real devil's work, which one has to approach through white magic to get to grips with."[28] Benjamin's project was not simply to demystify the occult elements of such authors as they often tapped into dimensions of experience neglected by rationalism, empirical science, and Enlightenment philosophy. Instead, Benjamin, as a proponent of the Romantic tradition of Marxism, sought the kernel of truth within occultism in order to redeem it through "white magic." In short, we take very seriously Michael Löwy's provocative claim that Benjamin's unique constellation of German Romanticism, Jewish messianism, and Marxism was itself an "alchemical fusion to produce philosopher's gold."[29]

We can also draw inspiration from Deleuze's engagement with the hermetic traditions. As Joshua Ramey argues, Deleuze's conceptions of being as an event and health as an ordeal of transformative becoming both bear a striking resemblance to the condensation, immolation, and reemergence characterizing the alchemical magnum opus.[30] Take for instance Deleuze and Guattari's reappraisal of the sorcerer in *Thousand Plateaus*, which, like Benjamin, finds new potential in an otherwise scandalized figure of anti-Enlightenment occultism. The sorcerer's powers lie in the ability to constantly produce becomings that cannot be directly represented and that alter existing power structures by introducing excessive forces and intensities into the world. Deleuze and Guattari refer to writers as "sorcerers" because they "experience the animal as the only population before which they are responsible."[31] This is not because the writer represents the interests of animals but rather because the writer enters into an entangled relationship with animals that enables the transversal exchange of affective intensities that can trespass species borders. The drift above and below species-specific characteristics cannot be verified or measured, and thus are occult (hidden) networks channeled through certain ritualized practices such as mantras, tantric states, hypnosis, trances, and possessions. The writer experiences animal magnitudes that relate seemingly unrelatable bodies through affect, speed, and colors, which all reside on the periphery of perception and common sense knowing. Writing resembles the magic of the sorcerer because it is an experiment predicated on "unnatural participation" with nonhuman others, cosmic intensities, and molecular vibrations that open up uncanny apertures in perception.[32] Much like alchemical monism, Deleuze and Guattari posit an "Interkingdom"[33] or cosmic plane that enables the exchange of forces and intensities across species that, once tapped into, can provide the creative powers needed to disrupt Oedipalized, dichotomous, hierarchical structures (epitomized by fascism). This is not to deny the destructive potential of sorcery (as Adorno would argue, and indeed, it is hard to deny the affective sorcery of the fascist agitator enabled by capitalist social technologies). Yet for Deleuze and Guattari, it is through such sorcery rather than despite it that radical trans-individual and trans-species experiments can produce qualitatively different ecologies of becoming.

And finally, we take inspiration from the book *Capitalist Sorcery: Breaking the Spell* by Phillipe Pignarre and Isabelle Stengers. Pignarre and Stengers define capitalism as "a system of sorcery without sorcerers (thinking of themselves as such)," that has "subjugated us primarily without the use of fire and sword."[34] The echoes of Marx and Engels are clear: As capitalism has advanced, it has acquired increasing power to intensify and extend itself beyond the control of the sorcerers who set it in motion. This peculiar form of sorcery is, for Pignarre and Stengers, based on the capacity of capitalism to capture the very capacities to live and to think. But the key point is that no one is outside this sorcery, not even those who might believe themselves to be safe (such as empirical scientists engaged in "objective" research in their laboratories). As such, a key political and aesthetic issue for Pignarre and Stengers concerns protection of the self and the commons from capitalist capture. Here, they reclaim the marginal/minor/defeated practice of "activist witches"[35] on the fringes of Western science and politics. For these authors, witchcraft has less to do with belief in the supernatural and more to do with a radically pragmatic experimentation in transforming what can be lived and thought. The recipes of witches are a "field for experimentation, cultivating the art of rituals able to give the inheritors of feminist, non-violent, anti-imperialist, ecological struggles the strength to resist the ordeal" of protecting themselves from capture.[36] Of particular importance is Pignarre and Stengers' emphasis on the witch circle as a *space and time* in which "the forces they have a vital need for can be convoked."[37] The circle is part of a pragmatics of witchcraft. It is not a laboratory that attempts to isolate and sanitize so as to objectively measure effects of carefully controlled forces. Instead, the circle is an invitation for anyone to participate who is willing to be "infected" or "poisoned" by a collective transformation in what can be seen, heard, tasted, and touched and thus resist the lure of capitalist sorcery with another kind of spellcraft. For us, casting a circle as described by Pignarre and Stengers will emerge as just one manifestation of the alchemical potentials of collective studioing practices, which are present whenever the dividing lines between science and art, head and hand, knowledge and pleasure are surreptitiously suspended in the name of collective experimentation.

We are intrigued by Benjamin and others and their constructive engagement with hermetic traditions. We join in their witchy circle in order to further the experiment in redeeming the potentialities that still remain (untapped perhaps) in the vast body of occult literature often dismissed by modern political, economic, and social systems (while at the same time parasitically channeling/siphoning off its powers). In particular, we are concerned with how the alchemical studio can present queer alternatives that are neither strictly regressive (fascistic and cultist) nor conservative (positivistic). Thus, we need to clear some ground and address possible worries up front.

A Historiological Worry

Any "historical" story we tell has to avoid two traps that are, in the end, inversions of one another. In the history of scientific development, the alchemical studio is often seen as a mere precursor to the chemistry laboratory, and is thus placed in a linear narrative of development in which science is read as a final cause or telos. This story erases what was unique about the alchemical studio to begin with, or reduces its novelty to some kind of obstacle needed to be overcome by advanced forms of scientific and more enlightened (and therefore more objective) knowledge. As Benjamin warns, there is a danger with historicism's insistence on narratives of progressive history in which the deficits of the past are negated and the future contains the promise of increasing improvements. This narrative construction amounts to nothing more than "empathy with the victor."[38] In other words, what might appear as historical continuity of progress is, in the last instance, a narrative written by the victors who use history to retroactively justify their actions as "good" all the while misrecognizing the unevenness of historical transformations. Other critics of the modernist notion of historical time include Latour who is skeptical of "time's arrow," which can only move forward by breaking with the past. This model denies what Latour sees as the disavowed temporality of mixture that always mediates "epochs, genres, and ideas" within a "collage" in which the "outdated and outmoded"[39] seem to (vestigially) live on. History, on this account, is more akin to a loop (more on this in the next gate!) in which "time becomes reversible instead of irreversible"[40] and the past becomes "revisited, repeated, surrounded, protected, recombined, reinterpreted, and reshuffled."[41] Or, we can think of Jane Bennett's potent criticisms of "disenchantment"[42] narratives of Western modernity in which reason triumphs over superstition, producing yet another linear narrative of progress that ultimately misrecognizes the forms of enchantment that remain, including the "affective alchemy"[43] of Walt Whitman's democratic poetry. In all cases, occult practices and esoteric beliefs residing within modernity continue to threaten the stability of the dichotomies between science and art, head and hand, knowledge and pleasure.[44]

Opposed to linear narratives of progressive disenchantment, we find value in Benjamin's experimentation with zones of ambiguity existing between the dis- and re-enchantment of modernity. Take for instance his formulation of the dialectical image in which the past explodes into the present and the present leaps into the past to create a shock of insight in which the most advanced practices and ideas are defamiliarized as premodern emblems and vice versa. He equates the dialectical image to "ball lighting,"[45] interrupting the chronologically linear narrativization of history with an abrupt and violent explosion of light (or a flash). Importantly, this ball lightning only appears to those who have cultivated a "seer's gaze" or "visionary gaze" that is untimely and imbued with a super-sensitivity (paranormal capacity?) toward the otherwise imperceptible "peaks of earlier generations as they sink further and further into the past."[46] The seer's gaze links dialectical images with the esoteric capacity to perceive that which is in excess of both biological, retinal vision and common, everyday, perceptual capacities. It pole-vaults the seer into another dimension that forms a "constellation"[47] out of exaggerated, far-flung elements that seem to lack intuitive connections yet provoke new insight into the present through a strategically unexpected abutment. For us, the constellation connecting alchemy to research-creation is one such dialectical image that will be explored in detail through each gate of this book.

The second historiographical error would be to simply call for a return to the alchemical studio as it really was. Such nostalgic longing for a lost past is dangerous on multiple levels, not to mention the (fascist) politics calling for a return to occult, Romantic, premodern modes of living. The approach would amount to nothing more than a narrative of continuity in reverse: The present is decadent or corrupted and thus a flight back to an unsoiled and pristine origin is necessary. This is a trap of romantic longing for an escape from modernity rather than the formation of a constellation (dialectical image) in which the present is shot through with explosive remnants of a past that are full of a tremendous potential for revolutionary leaps. The alchemical studio was a space and time for generative experimentation in research-creation, but it was also informed by highly problematic notions of perfectionism, hierarchical theories of the great chain of being, and could, at its worst, be seen as a space embodying extractive practices. All of this needs to be taken into account in order to prevent simplistic and naïve calls for neo-pagan revivals or new age spiritual rebirths informed by lost arts that somehow contain ancient "wisdom" and sacred "secrets" that harken us back to a more "authentic" mode of existence.[48] Again, for Benjamin, history is not a return to what was so much as a complex conjuring up of a lost potentiality *that never was* and yet continues to haunt the present, waiting to be fulfilled. For us, this potentiality resides in memoirs of alchemists, books of secrets, artistic imaginings, and allegorical depictions — all of which are shot through with untapped possibilities for queer experiments, anarchic speculations, fugitive dreams, and repressed desires that speak to a not-yet world to come (rather than a simple, neofascist return to what was).

Opposed to these historiographic tendencies (to progress or to return), we are opting for what we call a crypto-genealogy that redeems the alchemical potentiality of the studio for contemporary research-creation that, in all likelihood, was never actualized fully in any existing alchemist's studio (real or imaginary).[49] We thus engage in a kind of conjuring act in this book that concerns the idea of the studio, latent within historical manifestations yet not reducible to any one of them. This idea is a potentiality waiting to be developed in the present as an active intervention into the current split between art and science, studio and laboratory. Our selection of the term *crypt* for crypto-genealogy is taken directly from alchemical symbolism in which the alchemical vessel is often depicted as a grave or coffin in which the old king or primal material (the base matter connected to the earth that is necessary for the alchemical work to begin) is inserted in order to die and be reconstituted into the young prince or philosopher's stone. The crypt is thus the place of alchemical darkening or negredo, which is a necessary stage in the transmutation of elements. This is a space-time of potentialities or dark precursors that effect changes in state. Crypto-genealogy is interested in the question of accessing these potentialities that might be dormant yet be reanimated in the present moment.

Whereas genealogy might be thought of as *de*cryption, or an act of raising the dead out of the crypt into the light of reason, we instead plan to journey into the crypt itself, to see what remains underground, subterranean, and thus part of what Stefano Harney and Fred Moten might refer to as the undercommons, as a place of collective study.[50] At the same time, crypto-genealogy is not *en*cryption either, which merely privatizes what is in the crypt (think here of crypto-currency that attempts to commodify the crypt) or assigns cryptids some kind of ideological value within a system of signs. Decrypting and encrypting are locked into a binary that replicates the fundamental dichotomy between science and art, and as such, need to be problematized. At stake here is a much more obscure third possibility in which research (decryption) folds into art (encryption) and vice versa until the two become inoperative and thus are returned to a destituent or vestigial potential lost or abandoned in the crypt. Alchemical studioing happens *in the crypt*; it concerns *crypting* or the occupation of the crypt as a space-time of potentiality out of which decryption and encryption can happen. Studioing is cryptic work on knowledge that has fallen outside the division of labor between science and art, and as such, can only be traced through crypto-genealogical practices that *fabulate undercommon* constellations out of unfulfilled (destituent) potentialities that loop between past and present.

Besides Benjamin, perhaps the most apt precursor to our own crypto-genealogical method of research-creation is the long-forgotten father of paranormal writing Charles Fort. In his infamous *The Book of the Damned* (first published in 1919), Fort inaugurated a new form of writing that was neither simply scientific reporting nor poetry. As Ed Simon writes, Fort was not anti-science so much as critical of the positivism that had become science's dominant ideology, replacing the wide, indiscriminating, drifting nature of scientific curiosity with a much more controlled (and controlling) set of procedures (the scientific method), spaces (the laboratory), and criteria (verifiability and reproducibility) that, in the end, tended to foreclose more speculative approaches to inquiry.[51] In describing his own brand of research-creation, Fort writes, "I have gone into the outer darkness of scientific and philosophical transactions and proceedings, ultra-respectable, but covered with the dust of disregard. I have descended into journalism. I have come back with the quasi-souls of lost data."[52] Notice that Fort does not abandon science or philosophy. Rather he descends into the crypt (the outer darkness) in order to write new genealogies of that which is covered with dust out of neglect. Within his crypt, Fort compiled weird tales of exceptions that the scientific establishment would not dare to touch or which philosophical argument would find shameful or embarrassing. In the introduction to *The Book of the Damned*, Fort states, "A procession of the damned. By the damned, I mean the excluded. We shall have a procession of data that Science has excluded. Battalions of the accursed, captained by pallid data that I have exhumed, will march."[53] Crypts house exceptions — a theme that will return throughout this book and our engagements with Benjamin (states of exception), Alfred Jarry (laws of exceptions), and Georges Bataille (accursed share). In all cases, these theorists do not ignore exceptions, nor do they easily explain them away (through decryption and encryption processes that undermine or co-opt exceptions), but rather find in them the remnants of a potentiality to redraw the lines separating science from art, knowledge from pleasure, the head from the hand, the studio from the laboratory. As Fort argues, traveling into the crypt is destabilizing for science. He writes, "So it is, that having attempted to systematize, by ignoring externality to the greatest possible degree, the notion of things dropping in upon this earth, from externality, is as unsettling and unwelcome to Science as tin horns blowing is upon a musician's relatively systematic composition. . . ."[54] Science is not hospitable to intruders, opting instead to fortify barriers to things that might "drop in upon the earth" from the outside. Crypto-genealogy, as with Fort's writings, is ambiguous (full of perplexing riddles and strange connections that push us to the very edge of what is allowable to think or to sense — thinking as an act of transgression, out of bounds of the boundaries and barriers of common and good sense), and our book sticks with this lineage (even if it involves risks of awkwardness, unprofessionalism, or even madness) in order to thematize the perceptual, cognitive, and material states that induce and can be induced by exceptional drifts that swirl in dark corners of forgotten crypts. Fort provides us with a glimpse of an epistemology and a science open to the ambiguities of drift, allowing science and art to drift away from their institutional securities in order to mingle with less legitimate ways of knowing and being that do not attempt to damn up drift so much as embrace the drift of the damned. A new crypto-genealogical history is fabricated out of the damned.

A Worry About Sacrifice

Perhaps the most unique warning against a mere return to the alchemical studio (rather than a cryptic fabulation) is offered by Giorgio Agamben. For instance, in the chapter titled *"Opus Alchymicum"* from his book *The Fire and the Tale*,[55] Agamben outlines the dangers in any romantic nostalgia for the alchemical studio.[56] According to Agamben, the alchemical studio holds an important position in Western practices precisely because it complicates the science and art divide, and this inheritance was passed down through the Romantics, the Dadaists such as Duchamp, and culminated in certain pataphysical authors such as René Daumal and Raymond Roussel. The inheritance concerns the abolition of the work for the sake of a transformation of the self. The alchemical process to discover the philosopher's stone is, on Agamben's reading, actually an attempt to radically reconstruct the meaning of the life of the alchemist and the texts of alchemists chart this spiritual journey through the allegory of the transmutation of metals. This view of alchemy is perhaps best exemplified by the Victorian alchemist Mary Anne Atwood who fused alchemy with more contemporary interests in mesmerism. According to Atwood, ancient alchemy only *appeared* to be a physical science whereas in reality it was a science of the soul. Her starting point was an immaterial substance called "ether" — itself an updated version of the alchemical fifth essence, aether, or dynamic, unformed element/medium/energy that is underlying every change while also conserved within each change[57] — that was magnetically transmutated into the philosopher's stone not through physical experimentation but through a trance-like mental exercise. The philosopher's stone was not a material substance so much as the achievement of what Atwood described as a "higher plane of existence" or a state of self-purification.[58] This psychic interpretation of alchemy as a work on the self (and the self as a work) was also taken up by Jung who believed that the alchemist's unconscious psyche, and in particular the realm of dreams, were "projected" onto experimental materials. The allegorical symbolism of the alchemist's writings was not concealing secret knowledge so much as expressing the desires and dreams of the unconscious (and ultimately the collective unconscious).[59] Alchemy, in this sense, became therapy. This lesson lives on in new age magik in our present neoliberal society. Again, referring to Morrison's sigils cited earlier, the occult has become a practice of neoliberal, entrepreneurial self-fashioning or spiritual/monetary self-help/self-improvement.

Yet, there are decisive problems with this approach to the occult: without the production of a work, self-formation can transform into self-nihilation. Like Atwood, the pataphysician Alfred Jarry argued that a "luminiferous ether"[60] called *ethernity* opened up dimensions in reality that enabled the self to become increasingly flexible and discontinuous (in space and time). There is no doubt that this notion of ethernity was informed by Jarry's addiction to the colorless and volatile liquid known as ether. According to Jarry's biographer, Alastair Brotchie, his emphasis on inhaling large amounts of ether transformed pataphysics from an intensification of reality through the augmentation of actuality with virtuality to a science in the service of "radical individualism" and "pure subjectivity,"[61] ending in prolong stupors, anti-social withdrawal, and an untimely death. With the loss of the *opus*, so too vital insight supporting life was lost, and this loss was passed down through various avant-garde movements inspired by Jarry that turned away from works only to become absorbed into a self that, in the last instance, always verges on its own, narcissistic self-absorption and final destruction. In this sense, the absorption of the work into ethernity enacted by Jarry further exasperates the division between head and hand that is essential to capitalist production, ultimately eviscerating the hand and its works as completely vestigial. Stated differently, if alchemy refuses to sacrifice pleasure for knowledge and knowledge for pleasure, it nevertheless sacrifices the work (*opus*) for a subject that cannot, in the last instance, sustain itself on its spiritual path without annihilation.

W. B. Yeats also feared that alchemy demanded a sacrifice of the body, the work, and the world. On the one hand, Yeats was deeply committed to spiritual transformation through alchemical experimentation on the self, leading him to join the Hermetic Order of the Golden Dawn secret society. Yet he feared that the total devotion to the Great Work of alchemical transmutation of the self into a dematerialized, astral subjectivity might very well sacrifice the impulse to create works that work (renew the world). Summarizing this tension, Yeats wrote of his commitment to "nothing but states of mind, lyrical moments, intellectual essences" and of his desire to "hammer thoughts into unity" with regards to three important topics: the formation of literature, the formation of a philosophy, and a vision of Irish nationality.[62] While Jarry fell prey to the dangers of alchemical pataphysical experimentation on the self pushed to an absolute limit point of utterly dissolving, Yeats continued to struggle with the seemingly incompatible differences between radical transmutation and practical skepticism, philosophical prose and alchemical poetry, and worldly interests and transcendental interests without clear resolution.

For Agamben, there is a need to retain the work for exploring the unique potentialities of life, otherwise the experimentation inaugurated by the alchemist deteriorates into a solipsistic death wish. The head cannot, in this sense, lose contact with the hand without also sacrificing an essential potentiality that lies within a material experimentation with the world. Jarry is an extreme example of the latter as he virtually stopped writing and disappeared into a stupor of ether. The alchemist can only experiment on the potentiality of life through an act of research-creation that produces a work. But this is a strange work that is neither a completed work nor an absence of a work. In the same essay on alchemy, Agamben writes, "A truly poetic form of life is the one that contemplates in its opus its own potentiality to do and not to do, and finds peace in it. A living being can never be defined by its opus but only by its inoperativity, that is, by the way in which, maintaining itself, in an opus, in relation with pure potentiality, it constitutes itself as a form-of-life, in which what is at stake is no longer either life or opus but happiness."[63] At stake here is a form-of-life that is neither reducible to the works it produces nor abandons all works. In both cases, one's relationship to potentiality is totally lost. In terms of the former, the potentiality of life fully actualizes itself in the work, leaving no room to contemplate potentiality. The work absorbs potentiality into itself, into its actualization. In terms of the latter, the work ceases to act as a mediation point through which the alchemist can perceive potentiality, leading ultimately to a metaphorical or material act of self-destruction. Opposed to either option, the alchemist needs to establish a relay between work and self: The self contemplates its inoperativity to do and not to do in a work that reflects back this inoperative state. Because of this inoperative feedback loop between self and work, the alchemist can avoid sacrificing potentiality by continually rediscovering it, hence the rituality of many occult practices (including alchemy). The potentalization of potentiality through contemplating inoperative works (works that exhibit their unfulfilled potentiality) is the esoteric art of happiness for the alchemist. Said in an alchemical register, the occult search for gold (*aurum*) is distinct from the search for vulgar gold (*aurum vulgaris*). Whereas vulgar gold splits subject and object through the operative functioning of the market (reification and alienation), alchemical gold binds subject and object together in an inoperative moment of happiness (a poetic life).

The lesson here is clear: The alchemist's experiment must remain in the zone of indistinction between the *opus* and the subject without collapsing into either. This strange space is neither that of art (which the avant-gardes transformed into purely subjective enterprise) nor science (with its investment in the object). Instead, the inoperative work decompletes the subject, and the subject decompletes the work in the complex space and time of the studio. Or perhaps we can argue that the alchemical studio is the space-time of ethernity, which enables the drift between subject and object to happen — a contact point that neutralizes the strong poles of subject and object, rendering both inoperative before the sudden appearance of the potentiality of a form-of-life. If this is the case, then it is precisely the studio that is at stake in preventing the collapse of alchemy and pataphysical art movements into either pure interiority (self-therapy) or pure nihilation (self-sacrifice).

While thus far acting as a limit case, Agamben does praise Jarry for some of his works precisely because of their inoperativity. Agamben notes how he returns again and again in a loop to Jarry's writing as if the written word was itself a kind of pataphysical time machine in which linear, chronological time stops. These works, for Agamben, are inoperative insofar as they are "illegible." *Dr. Faustroll* is an occult work that "one cannot read" and is "outside of time."[64] Thus Jarry's alchemical notion of pataphysics necessitates a certain time and space in which subjects can lose their coordinates through engaging in producing and reading works that resist and yet nevertheless mysteriously fascinate them because they remain in potential (inexhaustible). These inoperative texts both demand continual interpretation and resist all interpretations at the same time, thus throwing into relief the potentiality to signify and not to signify simultaneously. The problem here is that Jarry ceased to engage with his own *opus*, leaving it for others to experiment with, in, and through their own studioing practices. While we might therefore think of Jarry as a kind of avant-garde/occult martyr, this itself would be too much of a romanticization. Instead, we can think of Jarry as the most extreme example of an internal problem with the occult: the temptation to retreat into a self that exhausts itself in itself, and thus can no longer engage with the potentiality to forge a new form-of-life found latent in the potentiality of its inoperative works.

Yeats encountered a different fate. Whereas Jarry dissolved himself, Yeats seems to have collapsed his alchemical longings into a political project with more than a hint of fascist aestheticization meant to impart form and direction to a decadent and wayward reality.[65] The attempt to put the alchemical dimension of his spiritualism to work in the name of political struggle and a search for strong Irish national identity led this conservative toward fascist sympathies. Missing here is recognition that alchemy demands a work on the self *through* the production of *inoperative works* that prefer to let idle such lofty aims and instead play with vestigial potentialities (useless to fascism). The strange appearance of such potentiality from within the ethernity of both alchemy and then pataphysics is precisely what exists betwixt and between alchemist and work, work and alchemist in a perpetual loop of indeterminant suspension. In both cases, the paradoxical spatial and temporal coordinates of studioing — and the inoperative works that both emerge from it and condition it — seem to be sacrificed.

Our repotentialization of alchemical studioing is therefore a space-time without sacrifice. Key to interrupting sacrifice is the experimentation with inoperativity and its unique connection to potentiality as an open-ended means (rather than operativity in which potentiality is a mere means to an end). We want to neutralize the sacrifice of knowledge for pleasure while also neutralizing the sacrifice of the work for the subject. At its best, studioing is a complex contact between subjects, practices (defined by procedural rules/constraints), and inoperative works. It is a space-time that knots together these elements, enabling them to pass through one another as a crucible for the invention of a new form-of-life that knows itself through inoperative works.

Romantic and Esoteric Worries

Third, we want to avoid any romanticization of the occult practices in the alchemical studio. Likewise, we ought to critically interrogate the meaning of occult "secrets" in order to make some distinctions. Throughout this book, we will engage with occult practices that might seem suspect (certainly not above Adorno's reproach), but not to glorify them or find within them some kind of premodern wisdom that must be rekindled (as with newagism or, even worse, fascistic cults with their secret rites and idol worship). At the crux of our analysis is the assumption that there is more to the occult than romantic nostalgia would let us believe. Like Benjamin, we want to deeply and creatively investigate the occult (and especially its spatial-temporal dimension, the alchemical studio) in order to render inoperative reactionary dimensions of it and release new potentialities, thereby inducing an alchemical transformation within alchemy itself that culminates in contemporary research-creation. To do so, the following book is an attempt to abide by the following rules. More particularly, we are interested in how these rules enable us to maintain the pliable space-time of studioing, which holds together in tenuous oscillation various movements, speeds, protocols, and spatial (dis)organizations (what we will describe later in terms of drift) without toppling over into *either* the laboratory *or* the artist's white cube, both of which are by-products of the split between head and hand, knowledge and pleasure, science and art that characterize so much of the modern, Western, capitalist world.

Rule 1: Embrace the Studious Secret

For instance, within alchemy there is a tradition of esotericism, which easily lends itself to cultish or clannish seclusion, privacy, and withdrawal from the world —, all of which speak to a radically anti-democratic, potentially elitist, and sectarian politics. But we are interested in how this very tradition of secrecy can be turned against itself to become an exoteric, and thus a common, practice of research-creation. Arthur Melzer argues that within mystical and philosophical traditions there are various forms of esotericism, each of which has its own set of motivations, styles, and degrees. At a minimum, Melzer defines the esoteric as the secret or hidden, which, he points out, was a rather common form of writing in both ancient and medieval worlds. Importantly, it was not always simply a trick meant to deceive.[66] Instead, esoteric writings often had political and/or pedagogical intentions. In terms of the former, such writings could convey hidden knowledge that might endanger the author if expressed openly and directly, and in terms of the latter, secretive writing could force audiences to think and discover for themselves. Already, we can see this at work within Zosimos's concept of the studious secret — a notion of the esoteric that is always already outward facing. Maier's early seventeenth century multimedia alchemical book *Atalanta fugiens* utilized allegorical emblems as pedagogical devices to draw readers into the study of alchemy by presenting sensorially simulating puzzles. The emblems made the study of alchemy entertaining and intellectually/spiritually alluring by simultaneously revealing and concealing secrets.[67] Studious secrets were not only written in books but also on the surfaces of buildings, transforming architecture into a text to be deciphered. The book *Dwellings of the Philosophers*, written by the mysterious author Fulcanelli, describes in great detail the alchemical emblems and allegories that decorate many medieval and early Renaissance French buildings, referring to such imagery as an "initiatory teaching"[68] into an art that is both secret and completely exposed for all to study. These architectural features thus withhold their teachings and give them as offerings at the very same time.

As Daniel Heller-Roazen observes,[69] secretive writing is paradoxical: It must be equal measure hidden and transmissible. Dark tongues, for Heller-Roazen, lie on the cusp between the esoteric and the exoteric. Riddles, which litter alchemical literature, exhibit two seemingly contradictory features simultaneously: (1) opacity and obscurity and (2) a hidden principle to their interpretation that is perceivable by the attentive reader. The riddle thus contains its own answer and as such, is continually giving itself away if one knows how to properly read esoteric texts. John Toland, in his work on esotericism from the eighteenth century, also points out that one should look for the key to a text within the text: "It is to be, for the most part, borrowed by the skillful from the writers themselves."[70] A studious secret demands an audience take it up, wrestle with it, ruminate over it, in order to study its *open secret*. In this sense, the esoteric dialectically becomes exoteric through the form of the studious secret. Collapsing any clear binary between the two, Antoine Faivre is quick to point out that not only is there an exoteric dimension to the esoteric (a tension between veiling and unveiling, hiding and revealing), but there is also an esoteric dimension to many purportedly exoteric texts, which always necessitates interpretation (and certain hermeneutic suspicions).[71]

In a similar vein, Deleuze has argued that esoteric knowledge is, in a queer way, democratic. Commenting on mathesis, which was a Romantic, vitalist science drawing on Hindu cosmology and numerology, Deleuze argues against common assumptions concerning hidden, occult practices. He writes, "to believe that mathesis is merely a mystical lore, inaccessible and superhuman, would be a complete mistake. This is the first misunderstanding of the word 'initiated' to be avoided."[72] The initiate is not someone who (a) has access to rarefied mystical lore or is (b) superhuman. Instead of a practice reserved for specially trained cultists, Deleuze argues that mathesis "deploys itself at the level of life, of living man,"[73] and is thus open and free for anyone at all to be an initiate. The non-knowledge of mathesis is esoteric in the sense that it cannot be directly represented or calculated, but this does not mean that it is withheld from the commonwealth (as if destined only for those who have private access and privilege). It is not private property at all but rather belongs to the living *as such*, without divisions or distinctions. For these reasons, the esoteric dimension of occult tradition can become radically democratic, giving itself away in order to exert the greatest possible transformative force on life in general. The space-time of the studious secret is therefore not as private and insular as one might at first assume, but rather opens itself up to the space of the commonwealth and to democratic dissensus over what can (and cannot) be sensed within its secrets.

Rule 2: Invite the Dark Precursor In

Although the occult is often synonymous with some kind of belief in spirits or aliens, we are not interested in the existence or inexistence of such entities. We are not making claims in this book concerning the nature of the proverbial "other" that haunts occultism. Indeed, the drive to "prove" the existence of aliens, ghosts, and various spirits, in a sense invests too much in the positivist sciences and their interest in measurement, verifiability, and prediction. Ironically enough, attempts to capture evidence of existence of various UFO phenomena ends up instrumentalizing otherness, attempting to control it so as to put it to work for pragmatic ends (capitalist or fascist). Likewise, new age emphasis on manifesting one's hopes and dreams through conjuring practices tends to erase the possibility of encountering the unknown excesses of experience, erasing dark supplements that defy the willful marching orders of a neoliberal, entrepreneurial self interested in bending the world to the will.

Opposed to these trends, we are more interested in how practices of studioing offer hospitable space-time for the arrival of otherness that exceeds subjective interests or positivist accounts of what "exists." The space-time generated in and through alchemical studioing makes room for forces, affects, signs, and entities that transcend and yet can be indirectly sensed by humans as queer swerves in the flavor of being. As Benjamin once observed, "Omens, presentiments, signals pass day and night through our organism like wave impulses" but can go unnoticed unless one cultivates sensorial sensitivities to these omens. For Benjamin, ". . . the only desirable telepathic miracle is a work of bodily presence of mind,"[74] which can be attuned to receive messages from alien, preconscious affective intensities coursing through our bodies that act as secret signals of possible, not-yet worlds of the future (or alternate dimensions of possibility locked within the present). What Benjamin is describing here is essentially an esoteric education of the sensorium to act as a radio receiver for "wave impulses" or innervations that can intensify and extend our capacities for new flavors of becoming.

For us, the studio is the space-time machine/apparatus for cultivating these sensitivities through the production of inoperative works. In this sense, the studio can be thought of as a hospitable place wherein Deleuze's "dark precursor" can suddenly appear out of the abyss in order to "set up communication" between "heterogeneous systems and even completely disparate things,"[75] and thus make surprising forms of creation possible that are unforeseen and unexpected. Indeed, Deleuze cites pataphysician Raymond Roussel's strange protocols as ways to tap into the excess of language as a kind of dark precursor allowing for the production of quasi-homonyms that lack any prior identity but nevertheless come to form a poetic series. The use of techniques, devices, and protocols for summoning the dark precursors (or as Benjamin would put it, uncanny omens, presentiments, and divinatory signals) is, as will be explored as we enter various gateways within the book, an important way to displace hegemonic forms of subjectivity (white, male, neurotypical, privileged ways of thinking and perceiving) and thus make place for some*thing* to take place that can disrupt the social, political, and economic spatial and temporal ordering of things. We cannot know what this something is or will be. In this sense, we appeal to the ancient writings of alchemist Zosimos who speaks of the philosopher's stone as that "which is not a stone, a precious thing which as no value, a thing of many shapes, this unknown which is known of all."[76] This is an object that cannot be known. Indeed, it is an impossible object, a true pataphysical exception, which prefers not to abide by the laws of scientific inquiry, the laws of logic, or even the laws of value. Indeed, the shock of the appearance of darkness is precious precisely because it lacks value, because it cannot be scientifically verified, because it lacks a clear function within the social, political, or economic order, because it is not reducible to subjective wants, desires, or hopes.

Rule 3: As Above, So Below

There is a perception of esoteric knowledge as inducing conspiratorial think-ing. Fascist paranoia is an intuitive sense that "everything is connected" and everything leads back to a master plot. Yet, the ability to produce similitudes across differences is a feature of occult practices that can be rehabilitated and is not reducible to these paranoid manifestations. Indeed, analogous thinking is so essential to the occult that this feature allows Brian Vickers to argue that astrology, alchemy, numerology, and natural magic are a "unified system."[77] In other words, the many guises of the occult each concern themselves with the discovery of resemblances between apparently unlike things or *discordia concors* (a series that connects through the intervention of a dark precursor). The cryptic symbolism of alchemy that includes various transactions between kings and queens, dragons and toads, eagles and hermaphrodites is a poetic strategy for juxtaposing opposites and bringing together far-flung things, revealing underlying universal interdependencies through correspondences (a tendency that also animates the strange aesthetic of Fort's paranormal writings discussed above). For Principe, an important occult factor defining alchemy is the belief in the interconnectedness of nature. He points to the sixteenth-century alchemist Tycho Brahe who referred to alchemy as "terrestrial astronomy" or "astronomy from below" and how the practice equated celestial bodies with certain earth-bound metals.[78] As a practical example, one can cite the famous alchemical experiment referred to as the "Tree of Diana" in which a chemical elixir of silver produced an object that resembled a tree bearing fruit. Through visual analogy, "Tree of Diana" "proved" deep, interconnectedness between macrocosm and microcosm, and organic and inorganic "life forms." Or again, the fourteenth-century text *Aura consurgens* compares the germination of the philosopher's stone with the nine-month development of the human embryo in the womb.[79]

As Benjamin might argue, alchemical writing was in touch with the primor-dial tendency toward recognizing sensual and nonsensual similarities through mimesis — a capacity that was alive in astronomy in which the movements of the heavens directly affected terrestrial events, thus putting humans in touch with inhuman, cosmic forces and intensities. For Benjamin, this mimetic faculty has an "increasing fragility"[80] in the modern, Western world precisely because of a shift away from occult sciences. For instance, in art, mimesis becomes disinterested contemplation, and in science, mimesis becomes detached objec-tivity. But perhaps most disturbing is the fascist "mimesis into death,"[81] which is the internalization of a negated, emptied world, creating the preconditions for self-sacrifice at the orders of the fascist leader. Mimesis into death is, we argue, also the final death of mimesis itself, or mimesis eating itself alive.

The fragmented, reified, abstracted world of capitalist technology often isolates or differentiates,[82] resulting in specialized spaces of inquiry (the split between the laboratory and the artist's studio). Insistence on division of labor in the sphere of production is internalized and replicated in relation to divisions between knowledge and pleasure, head and hand. Given the number of bifurcated and/or dispersed fields of knowledges and pleasures that no longer have a key that unifies them or can even provide points of articulation, the occult insistence on *discordia concors* is a powerful intervention into the present as is the space-time of analogical thinking that prefers not to abide by binaries but instead experiments with the open space betwixt and between (as above, so below; as outside, so within=a spatial twist that can form into ouroborotic knots). Or, perhaps even worse, in a world dominated by scientific reasoning, when sensual and non-sensual correspondences are posited, they are interpreted in causal ways. So much of conspiracy thinking, not to mention vulgar occultism circulating through new age discourses, is predicated on predicting events from distant and external features (such as stars) leading to fatalistic claims. But for Benjamin, producing mimetic resemblances across differences is not about immediate causal connections so much as about producing sign networks through which an incalculable (and thus dark) truth might be conjured (but never predicted or measured by science).

In strategies such as pataphysical syzygy, surrealist collage, Benjamin's constellational thinking, and Deleuze's combinatorials and diagrams, we find cryptic inheritances of the alchemical interest in the conjunction of things and the power of such conjunctions for igniting unexpected and surprising repetitions, loops, and repartitionings of the sensible (similarities that surprise across differences of space and time precisely because they have no preexisting identity shared between them). Indeed, Timothy Materer argues that the psychoanalytic idea of associationism between ideas as well as poetic devices such as metaphor and metonymy in modernist poetry link the psychoanalyst and the poet back to the shaman and link literary techniques with magic.[83] At stake here is overcoming the modern taboo against mimesis that was initiated by capitalist sciences (and suicidally deployed by fascism) in order to rediscover networks or series of relations that span great distances of time and space.[84]

Rule 4: Vitality!

As suggested previously, Bennett has recently argued for a reenchantment of matter.[85] Matter, rather than dead, inert, lifeless stuff waiting to be animated by human agency, has its own vibrancy that is not reducible to human-centered worlds. In "traditional materialism" there has been repeated emphasis placed on the need to return a distorted and reified materiality back to its explanatory origins in living, human labor, and the larger system of production of which it is a part. This line of critical analysis is best exemplified by Marx's description of the commodity as "stand[ing] on its head, and evolv[ing] out of its wooden brain grotesque ideas, far more wonderful than if it were to begin dancing of its own free will."[86] In other words, the appearance of the vibrancy of matter is generated by capitalism, which has a vested interest in masking the real origins of value. On this reading, the agentic view of matter would be a symptom of a problem: The enchantment of capitalism mystifies the human costs of production. The disenchanted world of capitalism is haunted by a more basic enchantment that is disavowed, therefore allowing the commodity to appear natural (rather than a conjuring of capitalist sorcery). Stated differently, capitalism hides the human behind the appearance of an autonomous thing (the commodity) that appears to have a life of its own. Arguing for the power of things would thus play into the hands of capitalist ideology and therefore be a bad abstraction. There is no doubt that this kind of ideological ploy makes the labor process invisible. At the same time, we find that traditional materialism has too quickly assumed that freedom from capitalist enchantment demands a turn back toward the head and hand of the human worker. As Bennett writes, "what demystification [*a la* traditional materialism] uncovers is always something human."[87] It is the human worker who must be emancipated from the enchantment of capitalist deception. But Bennett's new materialism offers a wider notion of emancipation: one that also sees the effects of the enchantment of commodities as exploitative of the preexisting agentic powers that adhere to material realities beyond the human. The dark powers of matter are beyond the human and speak to potentialities that are not reducible to human worlds or human ways of understanding.

But for us, there is nothing new about Bennett's theory. Indeed, new materialism's key insight is derivative of premodern alchemy and is therefore more traditional than traditional materialism. As historians have noted, ancient alchemy grew out of or in tandem with Iron Age metallurgy in which metals were conceived of as growing inside the earth as a subterranean womb. Instead of an origin in the desire to counterfeit gold or Greek scientific techniques, alchemy most likely grew out of the notion of the "Earth-Mother" (earth as belly, mines as wombs) bearing "embryo-ores"[88] that had their own vibrant life cycles. Indeed, artificially induced transmutation was probably modeled on the belief in what Mircea Eliade refers to as the "complex and dramatic Life of Matter" witnessed by miners, smelters, and smiths. Alchemy, in turn, was the knowledge given to smithies to help metals continue their growth toward perfection (in the form of gold and silver).[89] Fulcanelli summarizes this key alchemical point: "The vital activity, very apparent in animals and vegetables, is hardly less so in the mineral kingdom, although it requires the observer to pay more attention. Metals, in fact, are living and sensible bodies."[90] While humans might attempt to commodify the vibrant powers of the inorganic life of matter through increasingly sophisticated extractive measures, this ancient form of (in)organic life always exceeded such efforts as an overflow or overabundance that pulsates in the earth, in the darkness of caves, mines, crevasses, and tunnels — in the cryptic fecundity of metallurgic gestation.[91]

Of course, there is a danger here that such vitality will be perceived as an "essence" with an attending goal toward which growth orients itself. We see this in certain versions of alchemy in which gold is the evolutionary telos guiding the development of metals. Gold becomes the purification of fallen metals, which are now redeemed through transmutation. Vitalism becomes wedded to a purging of impurities, a distillation of life to a necessary essence (a single unifying elemental manifestation). A hierarchical division suddenly inserts itself into the flow of metals. This same esoteric narrative structure informs the contemporary archeofuturism of Guillaume Faye — a leader in the French New Right organization GRECE — who embraced the principle of "vitalistic constructivism" as inherently "anti-egalitarian."[92] Vitalism for Faye meant an embrace of forgotten, archaic values of the European essence — separation of gender roles, transmission of folk traditions, social hierarchies, a strong priestly caste, organic communities, the prestige of a warrior class, and so on — which gave Europe its will to power. On this view, egalitarianism is a threat to reclaiming the archaic precisely because it dilutes this essence. According to Faye's neofascist beliefs, "anarchism must cleanse futurism"[93] so that technology serves the interests of the European essence, enhancing its will to power. In fact, he argued for a de-democratization of techno-science so that it could become an "esoteric alchemy"[94] controlled by an elite class capable of protecting the European essence from the catastrophe of egalitarian politics.

In such cases, the essence/telos informing vitalism limits the potential of (in)organic life's becoming (resulting in capitalist extraction of value and fascist reactionary political formations). These "essences" are troubling insofar as they delineate what is possible in advance of any historical process. In response, we have to hold onto Bennett's key insight that vibrancy is an *empty* ontological principle (or perhaps we can say a *dark* ontological principle) that does not have a specific goal beyond its own experimental becoming. In this sense, "gold" would not be a measurable substance with specific properties (needing to be extracted from brute matter in order to generate value in the marketplace, or needing to purify itself of political and social contaminants in the fascist spectacle) so much as the vital principle manifesting itself through unexpected use/experimentation that is indifferent to ends. The vitalism of the "embryo-ores" would be unbound to market logics that insist on measuring value based on the "purity" of gold. Gold, on this reading, would become a latent potentiality that reminds us that any configuration of a substance can be *otherwise than* that which it appears to be. Such vitalism is vitally important for retaining sight of the key insight of alchemy: that all forms of matter are contingent, and that matter is always drifting away from itself.

Rule 5: Willingly Experiment with Profane Drifts

Finally, we are interested in the occult as a space-time of profane drift. By "profane," we refer to what Materer calls the occult's inherent "disreputableness" and its "rebelliousness"[95] and to what Fred Moten and Stefano Harney refer to as the fugitivity of black study in the undercommons.[96] "Occult status" indicates that something lies outside officially sanctioned canons of taste or strict boundaries of scientific knowledge. No canons here, only crypts. These occult practices press up against assumptions concerning legitimate pleasures and knowledges (and how the two might be distinguished from one another). This marginal status does not necessarily have to be viewed in a negative light (as indicating a failure to meet accepted standards). For us, the disreputableness of the occult can be reinterpreted as a virtue, or at least as containing a certain destituent and vestigial (useless) potentiality that makes it decisively non-functional for co-optation by fascist fanaticism or capitalist productivism. Indeed, when the occult becomes reputable and institutional it can quickly devolve into fascism. Perhaps there is even a definition here: Fascism happens when the occult is separated from its profane, inoperative dimension and becomes operative as the *sacred* (included into the homogenous order as a useful heterogenous surplus). As such, the profanity of drift is a necessary feature in order to maintain contact with the open and democratic space-time of an alchemical and pataphysical commonwealth.

Drift, in turn, can have several different meanings as will be explained in more detail in the second gate of this book. On the most basic level of material processes, alchemical experiments concern the transmutation of metals in the search for the gold of the philosopher. Mercury holds a privileged place in the alchemical imagination because of its strange, liquid-like properties. Matter is in movement, in various states of becoming. For Deleuze and Guattari, the occult practices and esoteric theories of alchemy are distinct from theology. Because theology focuses on defining "essential forms,"[97] there is no space and time for transmutations (so many becomings). Theology concerns itself with necessary and sufficient conditions for defining what something is, in its essence (so that it can be saved or damned). Opposed to this, Deleuze and Guattari turn to alchemy, which experiments with "accidental forms,"[98] or forms that are susceptible to variations in intensification. Alchemy subjects metals to intensities of heat and pressure, which enable various properties such as color, texture, and so on to interact with other intensities, forming composites that are "more or less"[99] golden. Alchemy deals with degrees that do not abide by the clear divisions and distinctions separating essential forms. This reading of alchemy releases it from the trappings of any teleology, and "perfection" of the metals becomes a contingent experiment in a substance's potentiality to be otherwise than itself (golden). Our concept of drift takes inspiration from the pivotal role of alchemy's mercurial principle, which emphasizes fluid change and transformation.

At the same time, we also highlight the importance of sulfur in alchemy. While mercury is the power to flow, sulfur represents the power to shape. One deforms while the other forms. The two are bound to one another in a dynamic and oscillating movement. Drift is not one or the other but rather their interplay, their contact. Drift is therefore not endless becoming, but rather can clump or mass into accidental forms necessary for the alchemist to contemplate (vestigial) potentiality, which has its own, immanent golden-like qualities outside the gold standard of the market. The contact between mercury and sulfur emphasizes the dual meanings of drift: to flow *and* to clump into accidental (inoperative) forms.

The idea of material drift and mutation was also incorporated into alchemical writing styles. As the ancient Arabic alchemist Jābir ibn-Ḥayyān once revealed, "my method is to present knowledge by cutting it up and dispersing it into many places."[100] Alchemical ideas drifted, dispersed themselves across texts, and it was up to the studious alchemist to read and assemble the parts into a constellation. What appears at first glance to be incomplete is really knowledge that points beyond itself toward drifting fragments (that may or may not ever be found, may or may not ever fit together easily). In this way, alchemy openly embraced the movement of drift, dispersion, and nomadic sprawl, and by doing so remained perpetually open, giving away its secrets to be studied. In fact, it was relatively common for esoteric authors to hide the meaning of their experiments by dispersing the meaning across multiple texts, which then had to be tracked down and carefully reassembled.[101] Yet any such attempt will never fully stabilize, and instead, the dispersed clues form a loose knot or mass (needing continual discernment) rather than a coherent system (as in laboratory sciences).

Epistemologically, drift enables us to avoid binaries between idealism and positivism and instead live at a point of indistinction between the two. In his writings on paranormal and esoteric phenomena, Fort once wrote, "We are not realists. We are not idealists. We are intermediatists. . . ."[102] Things drift, making it increasingly impossible to define and to know with any absolute certainty. Things are processes of becoming, of actualizing and deactualizing potentialities (and as such, nothing can be absolutely disregarded). This means that Fort cannot dismiss those epistemologies that are often characterized by good and common sense as illegitimate or profane. Something might sound outlandish, profaning the laws of good and common sense, yet it cannot be dismissed *prima facie*. An intermediatist can only create constellations between legitimate and illegitimate knowledges in order to cause swerves and drifts in what kinds of questions can be asked, what kinds of evidence can be collected, and what kinds of conclusions can be drawn (if any) through analogical juxtapositions. Fort's methods include those of "savages and scientists and children . . . cardinals and fortune tellers and evolutionists and peasants."[103] But this is not mere relativism for Fort. Instead, there emerges an epistemology of "gradations,"[104] or of drift in which identity/stability/definitiveness give way to states of intermediateness or becoming that no longer base judgment on either-or logic but rather embrace the creative propulsion of and . . . and . . . and. For Fort, the occult can generate drift that swerves the boundaries separating credulity and skepticism, undermining dogmatic fixity to one epistemological paradigm at the exclusion of others. Reality becomes a question of gradations.

Perhaps surprisingly, another prime example of epistemological drift can be found in the case of Sigmund Freud and his complex relationship with esoteric ideas. On the one hand, Freud dismissed esoteric writing as an expression of the "loss of value" in society after World War I and represented a return of "superseded convictions of primitive peoples."[105] In such exoteric, public quotes, Freud constructs a reassuring narrative of progress that places Western Europe at the pinnacle of reason. Yet, as Freud's biographer Ernest Jones recalls, Freud expressed both credulity and skepticism toward the occult. In his own esoteric letters and personal exchanges, Freud demonstrated a radical open-mindedness toward discourses and practices on the fringe of Western science. Thus, what is at stake here is actually epistemological attunement to what is on the fringes of what can be known (and enjoyed). For us, esoteric encounters induced a certain amount of drift into Freud's thinking that risked his professional standing (and even the legitimacy of psychoanalysis) while also opening up new possibilities for conceptualizing the nature of dreams, artistic creation, and the unconscious.[106] Interestingly, the imagist poet and patient of Freud, H. D., even referred to him as a modern "alchemist" who strengthened her beliefs in psychic powers and in the "alchemy of the word."[107] By extension, this would mean that the psychoanalytic chamber would be yet one more historical configuration of the alchemical studio. In short, an open mind is a mind that willingly gives itself over to a certain amount of epistemological drift.

Perceptions can also drift. Highlighting the ways in which the material experimentation of black artists induces swerves in what can be sensed, Key Jo Lee has coined the term "perceptual drift." Certain artists working within black feminist, new materialist dark ecologies break down distinctions and dichotomies between organic and inorganic material differences in order to explore the generative possibilities of (im)material vitalist powers to show that which would otherwise be occluded from the sensory field. In this sense, they embrace the enchantment of matters that have not yet come to matter within the commonsense ordering of things. Lee writes, perceptual drift "forefronts the multidirectional, multimodal, and polydimensional perspectival field necessary to plumb the occlusions and obscurities in the archive in order to narrate histories that attend to subjectivities and stories that are inconvenient to persistent master narratives because they make demands, assert refusals, and at times actually force us to move in our attempts to apprehend."[108] Lee emphasizes the force of drift. This is not an easy, leisurely waywardness, but a tug that strains what can be senses, that puts pressure on perception, snagging the eye or the ear toward that which would otherwise be overwritten by dominant dichotomies structuring experience.

The force of drift can intensify, bringing the very organs of perception to the edge of disorganization. At this point, we can also cite Deleuze's more violent notion of discordant harmony of the faculties when they are affected by a dark precursor. Whereas Kant attempted to fix the boundaries of the faculties, defining how each properly functions within certain limits (thus separating taste from reason), Deleuze is interested in radically open-ended, pataphysical and alchemical research through which "new faculties arise"[109] that do not yet have a function (and indeed, are characterized by their intensity rather than by a predetermined purpose). These new faculties do not abide by transcendental laws, but instead exist in a state of flux and tension. Here the swerve of the dark precursor generates such turbulent drift that the faculties themselves lose sense of what they are, how they function, and what their status is in relation to the mental division of labor outlined by Kant, producing new possible vectors for schizo desire to escape the gravitational pull of the Oedipal conflict. At a certain point, drift can transform into d-rift that pulls apart and reassembles the faculties. This is a more radical gestures than even Lee's emphasis on drift as unleashing a multidirectional, multimodal, and polydimensional perspectival field.

And finally, there is a connection between the visionary practices cultivated in the alchemical studio and neurodivergence as a heightened sensitivity toward the unseen in the seen, the unheard in the heard, and so forth. Both concern the sensation of excess on the perceiving body, and how the body can attune itself to forces, affects, and resonances of looping, twisting, drifting movements that would otherwise be undetected by neurotypical bodies. In this sense, there is a happy convergence of our own interest in crypting the space of the studio and the queer sensibilities of critical crip studies. As Manning argues, autistic perception is "the opening, in perception, to the uncategorized, to the unclassified"[110] and that is not yet characterized by hierarchical differentiation between senses, between human and nonhuman, between body and world. Autistic perception provides generative access to "a field perception that foregrounds the heterogeneity of a welling experience before it succumbs to the categorization of its parts."[111] Such perception is not about recognizing relations between subjects and objects so much as experimenting with contact resonances, and in this sense is the embodied, lived, sensorial enlivening/innervating of alchemical studioing. Drift, in all cases, is a disorienting swerve in what can be thought and what can be sensed that opens up to very different partitionings of the sensible that press up against (if not explode) common and good sense (neurotypicality). Drift indicates a pliability in the mind and the senses so that they can be stretched (opened) to wider vistas or subterranean flows, inflated to exaggerated proportions, and even burst open, allowing for new articulations of what bodies can do and what they can perceive. At stake here is a new notion of *neurodrift* as a movement of mental and perceptual faculties that can lead toward increasingly complex and intensified fields of experience.

In Melanie Yergeau's critique of intensive behavioral interventions for autistic and queer youth, they outline how this practice was (and continues to be) violent toward all things queer, effectively reducing the rhetorical possibilities of non-conforming bodies down to symptoms needing constant and vigilant rehabilitation. Importantly, Yergeau highlights how such therapies created relay points between the laboratory and the home. Yergeau writes, "The laboratory could not become the home — but the home conceivably could become the laboratory,"[112] meaning that the home could become a place of continual surveillance of queer bodies and family members could become co-therapists ready and willing to continually intervene if bodies deviated from their behavioral training. The expansion of the laboratory into the home increased the ability of science to monitor and correct both autistic and gender-variant children, and in a sense, the laboratory was an anti-drift technology of control. It is our contention that the alchemical studio follows another set of rules — rules that do not prohibit but rather experiment with what neurodrift makes possible as a form-of-life beyond neurotypicality, common sense rationality, and perceptual conformity.

Toward a Rational Occultism of Alchemical Studioing Practices for Research-Creation

In conclusion, by following these five rules, we see our project as an attempt to follow through on Benjamin's gesture and create a "rational occultism" that simultaneously recognizes the occult powers at work within reason and the reasonable dimensions of what might otherwise be dismissed as merely esoteric superstition.[113] Or, stated differently, rational occultism can be conceptualized as a profane contact zone between aesthetics and knowledge, art and science, head and hand. This concept enables us to be critical of postmodern occultism that dovetails with either capitalist entrepreneurialism or fascist mythology (both of which are irrational in their own ways). It also enables us to avoid the trap of positivistic empiricism, which regulates the occult to mere subjective hallucination or ideological illusion needing to be swept away through enlightenment reason and the controlled experiment of the laboratory. Such positivism misses how certain aspects of the occult are important reminders that creation happens through the sudden appearance of a dark precursor that is beyond the grasp of rational principles or instrumentalized and algorithmic calculations. *Rational occultism, or intermediatism, can become the queer epistemology of research-creation.*

In the rest of the book, we will turn to the question of the space and time of rational occultism: the alchemical studio. The dark drift of the studio is not merely the hybridization of what already exists but rather the creation of accidental forms of wayward, motely, and untethered modes of thinking, perceptual practices, bodily gestures, and materials that swerve into one another to form loops and knots (so many inoperative clumps). Transmutation is not merely transformation. Transformation might involve drift, but transmutation is a fundamental change in the internal structure of what something is, and thus is a d-rift or a rift in the process of becoming, a breakdown that is also a creative breakthrough of something else, something that does not abide by the laws defining the order of things (laws that concern essential being rather than accidental, wayward, queer becomings). The alchemical studio tenuously holds the inoperative works together (both subjects and objects now set a-drift from dominant binaries defining the modern, Western world), in a constellation, that does not sacrifice art for science or science for art, subject for object or object for subject, head for hand or hand for head. Instead, it openly experiments with what remains in potential in the space-time between such binaries: a dark potentiality that is not owned by anyone and thus gives itself away as the studious secret to an anticapitalist, antifascist, cryptic form-of-life.

Gate One

Dark Infrastructures of Studioing

At stake in our analysis is defining the space-time of research-creation. Because research-creation falls through the cracks of the academic division of labor that separates out the arts from the sciences, it exists in something of a limbo, or dark zone that does not abide by the laws of academic specialization. Through this gate, we posit that the studio of research-creation is informed by and forms around itself a web of *dark infrastructures*. This studio has an occult power to undo divisions between art and science, pleasure and knowing, head and hand, and is as disruptive as it is disreputable, hence the marginal (if not invisible) status of the space-time of research-creation in the academy. The darkness of the dark infrastructure of studioing practices takes on multiple valences of meaning ranging from its invisibility to its esoteric powers to make drift, make work, and make available as if by magic. Below we unpack these various meanings of darkness in order to ultimately return to the question of the alchemical studio as a paradigm of dark infrastructure. And in conclusion, we turn to two examples of how artists can reveal the witchy, occult powers of everyday infrastructures, in particular the humble fireplace and the window. This means that infrastructure is a contested terrain, and that the alchemy of research-creation does not necessitate its own specialized infrastructural organization so much as a subversive, fugitive pirating of what is already here, already infused with a vital, churning surplus or accursed darkness that is otherwise missed when infrastructure is reduced to supporting the management of everyday (academic, social, political, economic) life.

Defining the Darkness of Infrastructure

In the book *TechGnosis: Myth, Magic and Mysticism in the Age of Information*, Erik Davis traces our current fascination with and imagination of information technologies and infrastructures back to the ancient Greek stories of Hermes and by extension the hermetic traditions that bear the imprint of his name.[1] Indeed, the word *electricity* was derived from the writings of Jan Baptist van Helmont whose research into the healing properties of magnets positioned his experiments between science, art, and magic. Inspired by Helmont, the earliest books on electricity described it in alchemical terms such as "ethereal fire," which was passed down to Franklin, Tesla, and other inventers of the infrastructure of our information age. Summarizing, Davis writes, "Electricity, in particular, would carry three different aspects of the alchemical imagination into the modern world: the fascination with the vitality of bodies, the desire to spiritualize material forms, and the millenarian drive to transmute the energies of the earth into the divine realization of human dreams."[2] Mircea Eliade offers a similar argument. For Eliade, alchemy is not simply a precursor to modern chemistry, but is rather a mythical template for the quest for infinite energy, the acceleration of natural tempos of growth through chemical manipulation, and the generation of "immortal" synthetic products.[3] In this section, we take up the question of the infrastructure of alchemy and the alchemy of infrastructure, but from the subterrain crypts of the undercommons rather than from the perspective of physico-chemical sciences and capitalist industries, both of which have integrated the most productive and functional infrastructures of alchemical experimentation, such as the forge and the crucible, into the laboratory and the factory, respectively. In this sense, we recuperate a concept of *dark* infrastructure, or infrastructure that is animated by and embodies forces with occult powers and alchemical imaginaries that support research-creation from below.

Dark infrastructure brings together two concepts that have gained some recent currency in critical theory but rarely cross paths: infrastructure and darkness.[4] As anthropologist Brian Larkin has argued, infrastructure provides the grounds upon which objects and subjects operate without being reducible to these objects and subjects. It is literally the structure below the structures that support objects and subjects. "Infrastructures," writes Larkin, "are matter that enable the movement of other matter."[5] Infrastructures are material, yet are often displaced into the matter they move around. Because of this ambiguity, infrastructural substrates have a tendency to disappear, becoming increasingly invisible, or go dark. Opposed to this trend, Larkin calls for a return to the question of infrastructure *as such*, meaning a return to the question of infrastructural networks composed of technical, managerial, financial, and aesthetic structures that comprise the complex networks of our everyday daily lives. Such networks are not merely functional, but also concern translation across heterogeneous systems, routinization of certain (arbitrary) effects, and "ambient conditions of everyday life,"[6] including our collective sense of temperature, speed, florescence, and so forth. The diverse ways of understanding infrastructure all point to an underlying ambiguous potentiality within the concept, its power to elucidate facets of social, political, economic, and aesthetic life while also withdrawing or somehow receding from analytic clarity. Larkin frames this question in relation to methodological choice, but we would make a stronger claim: that the ontology of infrastructure itself contains a *dark* power that resists methodological thematization/operationalization.

Turning from anthropology to critical theory, we can begin to tease out the theme of infrastructural darkness. To begin, in *Anti-Oedipus: Capitalism and Schizophrenia*, Deleuze and Guattari argue that "desire belongs to the infrastructure, not to ideology: desire is in production as social production, just as production is in desire as desiring-production."[7] Here, desire is thought of as an impersonal, mechanic, productive force that is constitutive of the social field. Infrastructure, on this reading, is a set of dynamic, libidinal relations that can either stabilize into structures with concomitant ideological interests in *this party* or *this program* or can be set into dynamic, flexible, indeterminate drifts in order to maximize states of becoming and social experimentation. What is most important is that Deleuze and Guattari reroute the project of revolution from political and ideological concerns to infrastructural investments, speeds, and intensities. In this sense, infrastructure can be thought of as a plane of immanence upon which ideological structures or party formations might emerge to territorialize desire. For them, *infrastructural immanence* is a dynamic unfolding, a "groping experimentation" that is akin to a "witch's flight."[8] In passing, Deleuze and Guattari thus once again gesture toward a notion of infrastructure as an occult (witch's) power to produce movement and connectivity that somehow exist outside of scientific measurement or institutional oversight. Infrastructure exists in the twilight zone below or above articulated and concretized ideological frames and social formations.

Lauren Berlant argues that infrastructures are indeed ambivalent vectors of power, force, and movement. If infrastructure is a structure, it is a loose structure that is always in transition. They make an important distinction between infrastructure and institutions, writing "Institutions enclose and congeal power and interest and represent their legitimacy in the way they represent something reliable in the social, a predictability on which the social relies. Institutions norm reciprocity. What constitutes infrastructure in contrast are the patterns, habits, norms, a scene of assemblage and use."[9] Infrastructures can be appropriated or neglected, a space and time for differences to emerge, articulate, diverge, and enter into ambiguous contact with one another. For Berlant, infrastructure has a tendency to queer identity, likeness, and similitude in unintentional ways. In a close reading of the work of Juliana Spahr, Berlant extracts a notion of infrastructure as a "tensile structure . . . always making things different as they course through the material world."[10] This is a glitchy infrastructure that articulates multiple layers and levels of movements amongst and between mental, affective, and textual spaces, not always evenly, smoothly, or reliably but rather pierced by histories of colonialism, racism, patriarchy, and class exploitation that inevitably produce bumps, interruptions, and deep ambivalences. Importantly, Spahr's poetry is sparked by an "aspirational alchemical"[11] imaginary that bears witness to infrastructural patterns that risk a "common of awkwardness, complicity, and intimacy"[12] as a way to navigate the crisis ordinary of everyday life. Alchemy as a dark poetic art undoes claims of self-possession, common sense, and control, exposing the self and its relationship to the world to queer becomings and states of dispossession, dispersal, and distortion through infrastructural assemblages.

From a completely different philosophical lineage, critical theorist David Kishik (2015) highlights the dark connections between infrastructure and the occult. Kishik suggests that infrastructure is that which subtends and articulates the classical Marxian concepts of political superstructure and economic base.[13] It is, for Kishik, "anything that is understood in and of itself"[14] before it comes to serve economic or political functions. Infrastructure is a term that refers to things in their thus-ness, or how things are before this thus-ness is routed into serving specific ideological or economic imperatives. In relation to his theory of the metropolis, Kishik makes the passing comment that infrastructure is like a "city's id" the origin of which is "lying or lurking underneath both structure and superstructure."[15] Of course, the prefix "infra" means "below" or "beneath" but because of this, there is an implicit connection between infrastructure and darkness. It is that out of which *visible* economic and political life are derived but which remains in the dark, hidden beneath or behind or within these manifestations. Kishik's description enables us to further pivot toward the articulation of infrastructure with darkness. For all the X-raylike precision of critical theory, infrastructure has largely remained lying or lurking just out of view, always on the periphery of theoretical articulation—a witch's flight or desire/force that is unquantifiable in experience yet exerts a certain ambient intensity.

Keller Easterling offers another important point of reference for thinking about the occult darkness of infrastructure. For Easterling, infrastructure is more than physical networks. It is also composed of flows of information, actions, and energy. Differently from Kishik, Easterling argues "Far from hidden, infrastructure is often the overt point of contact and access, where the underlying rules of the world can be grasped in the space of everyday life."[16] While Kishik argues that infrastructure resides in the fuzzy background of critical theory (obscured by the more infamous dialectic existing between base and superstructure), for Easterling, it is perhaps even more mysterious still. Although readily apparent, an expanded notion of infrastructure as action and information remains nevertheless diffuse, intangible, and for all these reasons, *dark*. Take a simple example: the sidewalk. On the one hand, it is highly visible, yet on the other, it is totally cryptic. One walks *down* the sidewalk as if descending into a crypt or underworld. Our average, everyday language thus captures something of the darkness of the sidewalk, which its purported averageness seems to contradict. With the sidewalk there is an above below, which might also indicate a below above, and other strange spatial entanglements defining a host of infrastructural mediation points that mix-match spatial and temporal coordinates. Indeed, Easterling further describes infrastructure as a disposition or an immanent, indeterminate capacity that escapes explanation, "remain[ing] as a ghost"[17] within a script or technology yet never reducible to either. Drawing on Gilbert Ryle, Easterling concludes that this ghost in the machine acts as an "occult agency" existing in a "limbo world"[18] or what we might call a shadow world that exists below or beneath a certain threshold of perception while still exerting direct and indirect pressures on and through technologies, codes, scripts, and so forth.

In short, the witch's flight, the lurking, the pre-thematic thus-ness, and the ghost of infrastructure cumulate in Ryle's intimations of occult agency—all of which calls for an anticipatory alchemical imaginary (as hinted at by Berlant). This is the point of intersection between theories of infrastructure and the theme of (nonhuman) darkness. Our first inclination might be to think of this dark netherworld of strange, invisible infrastructural powers as full of danger and foreboding. Just as the "dark web" conjures up images of illicit transactions so too gentrification has come to symbolize the "dark side" of urban, infrastructural development. Or perhaps more ominous still is the darkness of platform capitalism, and the ominous retraction of the "do no evil" clause from Meta's mission statement. We might also think of top secret government agencies, hidden military bases, and covert laboratories that collectively make up the so called black world.[19] This perspective makes us pause and ask who are the secret agents behind the darkening of infrastructure? Yet the desiring-productive powers of Deleuze and Guattari's infrastructure, or the queer movements described by Berlant, or the ghostly web of occult agencies described by Easterling, or the thus-ness invoked by Kishik would seem to be beyond (or perhaps below) good and evil, and perhaps even express alien agencies beyond the control of corporate or government control. In short, darkness as infrastructure and infrastructural darkness both demand an ontological turn to that which subtends subjects, objects, and institutions (while also, inevitably, folding into these very same subjects, objects, and institutions).

For instance, on the most expansive, cosmological level, we can locate a swarm of nonhuman, dark infrastructures that persist above or below human-centered worlds. Some have speculated there is a haunting, occult presence of Dark Being that permanently withdraws (in the form of the abyss) from our common sense.[20] Or we might recall Deleuze's notion of the "dark precursor," which is "difference in itself or difference in the second degree which relates heterogeneous systems and even completely disparate things."[21] This is a darkness that does not withdraw but rather suddenly erupts into our perceptual and linguistic worlds with the force of the outside. Alternatively, François Laruelle speaks of a "transcendental darkness" that is "prior to light" and "escaped from the World before the World was born into the World."[22] In all cases, darkness demands a crypto-ontology that exceeds semiotic, social, and ideological registers not unlike the witchy, occult thus-ness of infrastructure. There is a dark infrastructure at the heart of the cosmos itself that is inaccessible to our known sciences and demands its own alchemical imagination capable of distilling new senses and flavors for dark matter and its opaque structures.

In sum, dark infrastructures operate through occult (hidden) powers that are sometimes material and sometimes immaterial. These forces, desires, agencies are the dark matters that enable matter to matter. The dark arts of alchemy appear to tap into such networks so that the most modest and human infrastructures suddenly touch cosmic forces that exceed intended, functional purposes. In this way, alchemy necessitates *and* produces complex relays of dark infrastructural networks that cross the gap between molecular and molar, historical and cosmological dichotomies. But if this is the case, then the question remains as to how to access, let alone depict, dark infrastructures.

The Problem of Representation

The problematics of infrastructure and darkness converge concerning the question of representation. As Michael Rubenstein, Bruce Robins, and Sophia Beal have pointed out, the question of "infrastructuralism" (that is, the study of infrastructure) revolves around making explicit various aesthetic strategies for defamiliarizing infrastructure just enough to make it appear in the foreground of our experience.[23] Without such strategies, infrastructure remains below, lurking in the background. While Christopher Breu and Jeffrey R. Di Leo offer an edited collection of essays on infrastructuralism that are, each in their own ways, "preoccupied with how *representation* intersects, shapes, and is shaped by infrastructure,"[24] we would add a secondary concern: how infrastructure is shaped by the representations that attempt to render it. To represent infrastructure is no innocent act insofar as it only appears when it ceases to function, in a state of breakdown. How then to represent that which is otherwise unrepresentable? How to resist the desire to turn the decisively silent running of infrastructure into a media spectacle that would betray the quintessential feature of that which it is trying to represent (infrastructure's darkness)?

Similarly in relation to the question of darkness, Darby English asks "How to see a work of art in total darkness?,"[25] especially when the techniques of art making embrace and exude such enveloping darkness? One possible answer to this question is the peculiar qualities of Kafka's writing, which Paul North describes as "beclouded" images.[26] For North, what is unique about Kafka is his ability to darken or muddy an image so that it no longer has its strong, positive attraction. Instead, the surface is merely rescinded without thereby revealing a hidden, deeper truth behind or beyond the surface. The mystery is that there is no mystery, or rather that there is only the mystery of the darkness always already on the surface of the image waiting to be released through acts of beclouding. Another answer might be found in Fred Moten's interpretation of Mondrian's post-optical, dark paintings as embodying "a para-ontology"[27] of the "incalculable rhythm of the life of things."[28] If dark infrastructure is paraontological—an incalculable rhythm that supports yet also escapes social ontology—then post-optical experiments are needed to "represent" the otherwise unrepresentable desires and witchy powers that flow through it.

In this way, mapping dark infrastructural networks would be an interruption of what Andrew Culp refers to as the "white infrastructure of photography."[29] Capture is, for Culp, the defining structural feature of such white infrastructure (think cybernetic codification), forcing darkness to reveal, expose, and perform for a white gaze. Darkness, on this reading, is an (an)aesthetic of invisibility (a beclouding), the refusal of the incalculable rhythms of the life of things to be fully present, and thus turned into spectacles of pain, suffering, or pathologized/instrumentalized otherness. In short, a new kind of (an)aesthetic process capable of supporting ways of seeing not predicated on a light-based aesthetic (or, at a minimum, light-based metaphors or analogies) must be conjured up. While the white infrastructure of photography attempts to illuminate darkness in order to police black bodies,[30] what is necessary is an (an)aesthetic that enables one to see in total darkness without illumination.

If we agree that an (an)aesthetic approach to dark infrastructure is called for, then we have to disagree with Levi Bryant's suggestion that "part of political practice would consist in diminishing the darkness of quasi-dark objects, of devising strategies to 'brighten' or intensify their appearance in situations."[31] Instead of such enlightenment aesthetics, which propose the diminution of darkness, we need more experimentation with how to bring *our* ways of sensing into the darkness. Instead of turning on the lights and illuminating the scene, Michel Serres instead opts to delve into the night and pay attention to its phenomenological qualities. He writes, "Night does not anaesthetize the skin, but makes it more subtly aware."[32] How can we endarken the senses so that they attune to shadows, fog, mist, and other modes of dark-becoming?

We can hear something of the question of accessing and representing dark infrastructures echoing throughout Édouard Glissant's notion of the *right to opacity* for oppressed peoples who have experienced the demand to be seen and represented as a form of (white, cybernetic) power. For Glissant, "That which protects the Diverse we call opacity."[33] Anne Dufourmantelle states it best when she asks how we might "gaze on something in darkness without any possibility of being illuminated. . . ."[34] This does not mean the gaze that stares into darkness's opacity is incapable of seeing. Indeed, for Dufourmantelle, intimacy with such darkness makes us into "seers"[35] beyond what is given to the senses in everyday perception. The seer can *see* not despite of the darkness but because of the darkness. Seers see within darkness a plenitude that might otherwise be missed—suddenly infrastructural webbings pulse with incalculable rhythms. Seers, on this reading, are those who have cultivated post-optical (an)aesthetic strategies for discovering, tapping into, and channeling the witchy powers of dark infrastructures without reducing such infrastructures to the kinds of calculation found in white cybernetic capture.

But perhaps one more step is needed here. It is not simply the darkness of dark infrastructure that makes it so difficult to render, it is rather the *blackness* of the darkness that is at stake here. Eugene Thacker's genealogy of darkness, blackness, and night in *Starry Speculative Corpse* provides a short overview of various mystical, philosophical, and aesthetic approaches to thinking through dark arts. In particular, he draws upon Nicola Masciandaro's argument that darkness is a property of blackness while blackness is not reducible to darkness. Thacker then speculates: "Perhaps there is a black that is seen—a black of shading and gradients—as well as a black that is unseen—the black of retinal inactivity."[36] On our reading, the blackness that can be seen is more properly darkness whereas the blackness that cannot be seen and can only be accessed through a seer's unique capacities for post-optical (an)aesthetic sensation is *blackness*. This distinction gives specificity to the blackness on the periphery or limits of sensation. Without darkness, blackness is simply the inaccessible thing-in-itself, and without blackness, darkness is simply a by-product of white infrastructures of surveillance and governance. At stake here is an (an)aesthetic capable of a radicalization of darkness (total darkness = blackness = blackout) to the point where its dependence on light suddenly transgresses its own limits.

Cultivating dark perceptual capacities is particularly difficult given the temporality of blackness. Whereas light always comes from a source and thus has a beginning (an earlier and later), blackness does not have such an origin, and thus appears outside of temporality. Light projects out, and for this reason, has a history to it. Yet blackness is not a projection. Instead, it merely subtends in its thus-ness, or we might say it withdraws into its thus-ness. If there is any way to figure such blackness it might be through the holes it leaves within light (the darkness of shadows, for instance). Dark infrastructure, on this reading, would appear (if at all) only in the moments of temporal break or interruption when the blackness of darkness produces a hole in the surrounding lightscape. Stated differently, we can only trace the shadows of dark infrastructure, which themselves are tracings of an incalculable rhythm—a witchy power—that is cosmic. This is precisely why the witchy powers of the lurking darkness within infrastructure cannot be fully operationalized. Indeed, at the very heart of the darkness of infrastructure is an inoperative core of blackness that prefers not to be calculated.

Dark Traces

We can discover ontical traces of darkness at work in subterranean political, technological, educational, and ecological infrastructures that support acts of fugitivity and anonymity. For instance, Culp calls for a conspiratorial communism that suspends the integrating logics of cybernetic control. Whereas cybernetic policing functions through the logic of visibility and representation, Culp highlights those guerilla-like insurgency movements that refuse to partake in the politics of recognition. Instead, these ungovernable subalterns are guided by a philosophy of subtraction, anonymity, and imperceptibility. Drawing on Deleuze and Guattari, Culp describes the imperceptible as "that which interrupts simulations, sort-circuits diagrams, and evades representations . . . . receding before it can be captured, and put to work for the benefit of the state."[37] As an example, Culp highlights the distinction between armed militants (who merely imitated the violence of the state) and more creative, intensive guerrilla movements that emerged out of Italy's Years of Lead (late 1960s until the late 1980s). The latter experimented with subterranean radio stations, bands, celebration, riots, and squats all of which formed what Culp refers to as "an empty architecture of indistinction, informality, and semisecrecy."[38] Here, Culp emphasizes the subversive use of abandoned and/or marginal infrastructural elements and powers for experimenting with antifascist, anarchic potentialities of escape. Extrapolating from the radical Italian politics of escape, Culp argues that the new guerrilla must become "a scholar of density,"[39] capable of finding exit routes in and out of the cybernetic metropolis so as to intensify the flexibility and mobility of insurgents. "There are many parts of the new cybernetic landscape that appear as dark as a moonless night even when the sun is shining its brightest, for anonymity is to our contemporary metropolis as the cover of nighttime is to the city."[40] In this analysis, Culp points out the importance of dark (counter)infrastructural networks simultaneously enabling diffusion/

withdrawal of insurrectary actors while also enabling them to remain loosely connected (without a centralized head, organized party, or elected leader to unite them). Another example of the politics of dark infrastructure can be found in Ezekiel Dixon-Román and Ramon Amaro's notion of "Black techno-conjuring,"[41] which not only diagnoses the racial and colonialist violence of white cybernetics but also the anti-colonialist hauntings of such digital infrastructures by the creative indeterminacies of blackness. The thus-ness of infrastructure might therefore lend itself to undoings, exits, and tactical subversions of "recursive colonialisms"[42] that can remain imperceptible to state control or capitalist expropriation.

Against the prevailing common sense assumption that schools should enlighten (bring into the light of the truth), educational theorists are increasingly calling for an "endarkenment" of educational philosophy.[43] Derek Ford and Tyson E. Lewis[44] argue that the darkness of opacity (a refusal to articulate and explicate) is an important aspect of communist pedagogy of escape while Harney and Moten[45] have theorized the undercommons as a form of dark infrastructure, or a place of fugitive, black study that refuses to be policed, controlled, or managed by the audit culture of educational institutions. In both cases, the fugitive nature of study is emphasized as a way to resist the white cybernetic emphasis on calculation, visibility, and representation that are so prevalent in higher education. The informal networks of outcasts, dropouts/pushouts, and political dissidents that study in the undercommons replace economics of debt with relational or social debt. Perhaps we can even go so far as to suggest that the opacity of the undercommons enacts an exorcism of bad debt from educational practices, leaving open a gap for a dim (nearly imperceptible) educational form of life beyond capture by learning metrics, assessment measures, and analytics of excellence. Such would be a mysterious (if not wondrously weird) education. At stake in education is a dialectic of darkness comprised of the militarization and corporatization of the academe[46] and another form of excessive darkness that generates shadows fostering undercommons and alchemical experimentations against forces of control.

The cosmic, nonhuman traces of dark infrastructure can also be found lurking in the shadows of Timothy Morton's notion of the "strange loop" of dark ecology.[47] For Morton, ecological theory must attempt to think the weird and uncanny dimensions of appearances beyond mechanical, causal understandings of "nature." Drawing on the ancient Norse meaning of weird as twisted, Morton proposes ecological awareness as awareness of the strange loops in which "two levels that appear utterly separate flip into one another,"[48] such as the weird way in which nature and culture, individual and species, geology and humanity contact one another with a sudden, uncanny urgency. One might think of infrastructure in terms of the negative feedback loop of "agrologistics"[49] as a technical, planned approach to global agriculture, but what we are emphasizing is another counter-infrastructure constituting the weird, mysterious, wonderous feedback loops of dark ecology. On our reading, Morton is attempting to theorize (and in some sense poetically enact) the invisible and inaccessible infrastructural loop of witchy, occult powers that articulates otherwise distinct levels, dimensions, strata of an unbound paraontology of dark becomings.

Studioing in the Dark/as the Dark

The traces of dark infrastructure are everywhere. There is a sense of darkness emanating from above in the form of surveillance cameras, drones, and the likes, while also a subversive, fugitive darkness from an infrastructural under-commons that composes new space-time configurations from below. While the former can be considered the appropriation of darkness (denuded of its vital relation to blackness), the latter can be considered a point of resistance (an embrace of the blackness of darkness). In conclusion, we want to highlight how the alchemy of the space-time of the studio is perhaps the prime example of the political, educational, (an)aesthetic, and ecological dimensions of dark infrastructure from below. This despite the fact that the studio has undergone multiple campaigns to literally and figuratively whiten it. One can think of the movement toward the modernist notion of the studio as a "white cube" sani-tized and purified of all contaminants and impurities.[50] Take for instance the factory-style "laboratory workshops" of the Bauhaus. Or, as modernist archi-tect Le Corbusier once wrote,

> Every citizen is required to replace his hangings, his damasks,
> his wall-papers, his stencils, with a plain coat of white ripolin.
> *His home* is made clean. There are no more dirty, dark corners.
> *Everything is shown as it is*. Then comes *inner* cleanness, for
> the course adopted leads to refusal to allow anything at all
> which is not correct, authorized, intended, desired, thought-out:
> no action before thought. When you are surrounded with shad-
> ows and dark corners you are at home only as far as the hazy
> edges of the darkness your eyes cannot penetrate. You are not
> master in your own house. Once you have put ripolin on your
> walls you will be *master of yourself*.[51]

The so-called "Law of Ripolin" was concocted in 1925 for male property own-ers who were required by Le Corbusier to paint their walls white. The sanitized white cube was a sign of enlightened hygiene, clarity, truth, purity, and, most importantly, mastery over one's self and one's space (the supreme form of sur-veillance and governance). To master space was, for Le Corbusier, to master the self, as if the darkness of the external space of the white cube was directly connected to the darkness of one's unconscious (polluted by filthy desires or the nonknowledge of excessive otherness). Everything must be shown as it is, without secrets. This is a radical exteriorization, an absolute visibility, an archi-tectural form of white cybernetics. Indeed, for Le Corbusier, one cannot be "at home" as long as there are shadows. Any trace of lingering darkness must be abolished in order to establish a home (but a home for whom?). Rooms must abide by a strict law so that the dangers of darkness do not creep into one's life. The law cannot be deviated from. This is a straight space, or a space that must rigorously straighten itself out. Le Corbusier therefore offers a radical critique of the darkness of infrastructure. There are simply too many risks in inhabiting and working within dark spaces. These risks compromise (white, male) mastery. All darkness, all esoteric secrets are reduced to deviances abolished by the Law of Ripolin. This is a space inhospitable to the anomalous arrival of a dark precursor, and by extension, cannot abide by any intrusion that is not sanctioned in advance by the master of the studio space.

We can juxtapose Le Corbusier's paranoid, light-filled space with the "dark space" theorized by Eugène Minkowski. Whereas Le Corbusier's space individuates through disciplinary, panoptic surveillance, Minkowski's dark space depersonalizes and disappears. The living space of darkness absorbs the individual to the point where it is almost impossible to tell the difference between the individual and the deep, dark milieu. In this sense, darkness is not simply void but is a radical fullness that "touches the individual directly, envelops him, penetrates him, and even passes through him."[52] The permeability of the ego is experienced in darkness rather than in the light. The subject becomes dark, even to him/herself. The origin of the fear of darkness (Le Corbusier) is found in the "magical" "ascendancy of night and of the dark,"[53] which poses a threat to the integrity of the organism, leading to the concomitant desire to enlighten all dark spaces through white cybernetics. As Dufourmantelle argues, the magic of darkness is a risk. The night, as a phenomenological envelop of darkness, is "our secret amplitude," which can only be envisioned (or endarkened) through enigmas, nonsense, and visions that "dismantle us from within."[54] The night, on this view, can strip away recognition, functionality, and commodifications, leaving behind a naked, becoming-imperceptible. At the edge of darkness, you risk "undoing yourself."[55] We agree with Dufourmantelle, but we want to specify the "us" and the "you" who are undone. Within a world dominated by white cybernetic surveillance and governance, the mastery of the white, male, property owning subject (the subject of the home, the subject who is supposed to be at home) is precisely what is at stake in any encounter with darkness, hence Le Corbusier's phobic paranoia.

A crypto-genealogical fabulation of the alchemical studio emphasizes and champions precisely those qualities of darkness, which Le Corbusier is paranoid of. For us, alchemical studioing is a fugitive space-time machine, which concerns endarkenment rather than enlightenment and proposes a new ecology of perceiving bodies and dim, barely legible, inoperative works. It cannot be built according to any physical laws or rational plans. Instead, it has to emerge through a darkening process, a witch's flight that is populated by a swarm of unnatural figures including werewolves, sorcerers, alchemists, and hobos (as well as artists and scholars and scientists, but who can tell the difference down here in the darkness where things drift astray?). This is a space that *lacks mastery* (remember those awkward alchemists), that embraces dark corners and shadows in order to be hospitable to the arrival of dark precursors that fundamentally challenge what can be seen, heard, sensed, thought. Here darkness is not indicative of illness or corruption but rather a necessary feature of studio infrastructure, making possible swerves and queer trajectories that rupture the rigid straightjacket of the white cube described above. It is not law-abiding but is rather anarchistic and thus *never a home* but rather a machine for perpetual experimentation with gestures, bodies, thoughts, and intensities beyond rational, logical, measured, enlightened sciences. And in this way, the darkness of studioing offers an *impossible solution* to the current problem of research-creation—a problem for which the practical solutions of the laboratory cannot address (let alone perceive).

In the darkness of the studio, alchemy happens. At the same time, alchemy produces darkness. It does not reduce darkness to an effect of the light, but discovers that its many shades emanate from blackness as an incalculable surplus of vital forces and desires. Only post-optical seers who are attuned to the drift of witchy powers can find their way through the gate into the studio, and once there, only they can activate, catalyze, and bear witness and be affected by the new tastes, pleasures, and flavors from the blackness.

The studio as dark infrastructure for alchemical experimentation with more-than-human becomings is eloquently described in *The Book of the Art* by the Florentine painter Cennino Cennini. The dark infrastructure of the studio allowed Cennini to "find invisible things hiding in the shadow of ones in nature" such as a figure of a "half-man, half-horse."[56] The studio supported (and was supported by) mental and perceptual drift that conjured up a posthuman figure, an inhuman figure from within the fecund shadows of nature. In what follows, we will turn to two examples of dark infrastructure—the fireplace and the window—in two very different works of art. What these works reveal is the ultimate secret of research-creation: that the darkness of dark infrastructure is hidden in plain sight on a post-optical field of sensation. All that it takes to tap into the occult, witchy potentiality of such infrastructure is a dimming down of the gaze and a sensitivity to the "invisible things hiding" in the shadows.

The Fireplace

Studioing has always been associated with the night. As an engraving of Baccio Bandinelli's "Night Class" from mid-sixteenth century illustrates (figure 1), the nighttime studio was potentially a collective, convivial affair. The image depicts a bustling, almost chaotic scene of figures engaged in a number of activities: drawing by candlelight, reading, ruminating, discussing, experimenting. The light from the fire and the candles takes on an intensity and vibrancy that transforms them into stars as if the heavens themselves have come down to earth and entered into this nocturnal cave-like space. The room is full of replicas of Greek statues, books, but also various anatomical fragments scattered across the floor. In his analysis of this strange depiction of nocturnal studioing, James Hall writes, "We could be eavesdropping on a wizard's coven or alchemist's lair."[57] For Hall, beyond the fact that studioing takes place at night, there is something "dark" about the scene as such that aligns it with disreputable practices such as witchcraft and alchemy. Interestingly, Hall points out that the civilizing and legitimizing term "Academia" is not inscribed on the print, thus indicating something fugitive about the darkness of the practice—it is a space and time a-part of/from any legitimate institution.

While the light of the candle might indicate enlightened study, it is important to note that the candles produce ominous, elongated, deformed shadows that are cast over many of the depicted faces. Thus, the light does not eliminate the shadows but produces them, and it is within the shadows rather than against them that studioing practices can take place. The shadows are thus not incidental features of the studio but are at the very essence of the studio. They are part of the *dark infrastructure* of the studio, supporting in an immaterial almost ghostly way the production of images and ideas outside of the official academic curriculum. As Juhani Pallasmaa writes, "deep shadows" "dim the sharpness of vision, make depth and distance ambiguous and invite unconscious peripheral

Figure 1. Engraving by Ene Enea Vico based on Baccio Bandinelli's "The Night Class" (c. 1545–1550). Source: Public domain.

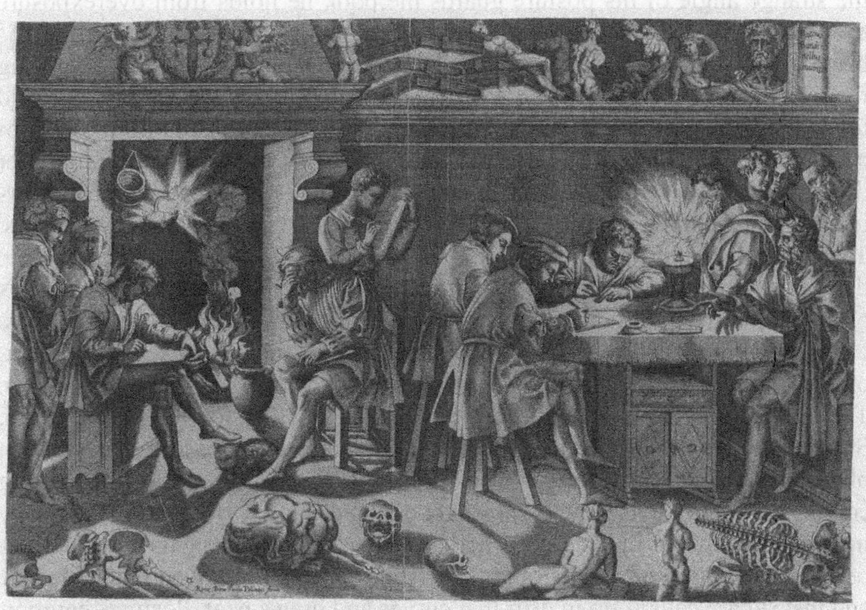

vision and tactile fantasy."[58] Visual senses have to be diluted and the visual field has to be rendered murky, ambiguous, and misty for some*thing* to emerge on the periphery of what can be sensed. The shadows thus support an endarkened form of education that is not predicated on illumination (and attending surveillance) so much as a sensorial drift that opens the body up to dark matters on the periphery of known worlds.

Within the murkiness of the darkness, the statues and skeletal remains almost seem to take on lives of their own. They are, in other words, reanimated by the enveloping darkness as it consumes the light of the candles and the fire. The objects appear animated by an occult power or an atmospheric charge that can bring to life what would otherwise be inert or dead. Instead of residing peacefully in their proper places according to a specific, taxonomic order or according to an educational curriculum, the objects are scattered (seemingly haphazardly), almost as if they have gotten up and *moved themselves*, crawling across the floor. The dark infrastructure of the studio has come to life in order to reveal its powers to shape what happens inside its chamber, how things appear, what can move, and how bodies are arranged. The image is therefore not, as Hall suggests, merely a conventional depiction of the studio as a "ruinscape"[59] (a crypt to house dead body parts) but is actually a dark ecology of encroaching shadows, animated objects, flickering candles, and human practices that collectively twist death into life, light into darkness, and so forth.

Needed here is adaptation of the eye to dim conditions in which weird and witchy visitations can happen. The light is not to bring to light so much as to produce shadows. Indeed, the figure of Bandinelli in the center of the image sits near a fire, but, in an important gesture, shields his eyes from its light, and the strange angle of his posture guards the book he holds from overexposure to the fire. It is as if his vision has become increasingly light sensitive. His back is also curved toward the book. His body posture is not straight, gazing forward toward a horizon line; instead, his posture is decisively bent downward toward shadows, which certainly carries moral implications. Does the lack of correct, upward, vertical posture indicate a lack of moral will and thus queer deviancy? Likewise, the figure behind Bandinelli studies an anatomical drawing, yet instead of holding the drawing into the light, he holds into the shadows. To see, he does not turn to the light but away from the light. One might argue that this gesture is to ensure the light of the fire is cast upon the text, but his own shadow would preclude such a reading. Furthermore, he draws the image close to his eye as if he is intentionally sensitizing his eye to see in the dark, or even more tellingly, the double darkness of the room and his own shadow. The text is not therefore illuminated but dimmed by this doubling of darkness. Such a practice demands a heightened intimacy between the individual and an object that would not be necessary if the object were viewed in full lighting conditions. In this sense, the night studio demands new ways of seeing and new intimacies with the objects (and perhaps ideas) that are engaged with. Both bodies have fallen off of a straight, vertical, detached/removed/objective line and instead are caught up in drift that curves and draws near. Postures comport to the demands of darkness, transforming the shapes and gestures of bodies. Studioing is not only the manipulation of objects but also concerns how these objects (and their occult powers) shape the body—drawing the eyes near or creating a curvature in the spine as the body bends forward to read. Body and object enter into ouroborotic loops generated through studioing. Sight sensitizes to a becoming-imperceptible, and the body transfigures itself into a curved and twisted nocturnal form.

On our reading, Bandinelli's studio is a depiction of the conditions necessary to invite the dark precursor and thus stands as a crypto-genealogical antidote to Le Corbusier's white cube. It is, in a sense, a portal or opening for something to appear, or rather make an appearance by remaining imperceptible. Low lighting conditions of dark infrastructure are hospitable because they do not insist that the dark precursor show itself and thus can remain enigmatic, opaque, and fugitive. Bandinelli's studio is decisively not a technology of white cybernetics as discussed above. Significantly, the fireplace is nearly pitch black—a hole puncturing the image right at the point at which it should be most illuminated. No light from the fire is cast upon its back wall, or perhaps there is no wall there at all but a gate to another room (or dimension, subterranean or cosmic or *both*). While the fireplace might, at first, appear to lack all spatial depth, it can also be read as an opaque and unlimited space. The fireplace becomes a gateway to a dark cosmos—a gateway that does not lead outside the space-time of the studio but rather to the abyssal, internal, radical blackness (paraontological blackness as Moten would say) at the heart of the studio . . . the heart of its darkness. Hanging in the void of the fireplace is an alchemical star as if the night sky itself existed within the studio. As above, so below. The heavenly night sky sinks into the crypt producing an excess or expenditure that travels in and out of perception, both absorbing the light around it while also emitting its own dark matter. What appears at first to be nothing more than an empty fireplace is actually a post-representational, (an)aesthetic depiction of what cannot be depicted as the very heart of studioing around which all other activities, practices, and rituals revolve: a cove for sheltering the lurking thusness of dark infrastructure within the studio.

But, of course, it is important to point out that all the participants in this scene are white men. One reading would be that the studio and studioing practices are the domain of men (circumscribed by the patriarchal violence of exclusion and exclusivity). This is certainly the historically accurate reading of studioing images. And yet we want to pose a crypto-genealogical reading that takes seriously Hall's connection between this image and the alchemical studio. The deformations of the male body indicate how their whiteness twists in the presence of darkness and how their eyes must adapt in order to see in the shadows. So perhaps the studio is not the space-time for sheltering whiteness but rather of deforming whiteness under the pressure of a dark precursor emitted from the blackness of ethernity that drifts in and out of the pataphysical portal at the heart of every studioing practice. The darkness of dark infrastructure does not submit to whiteness but rather acts upon it, and in this sense, whiteness must exert extra effort to straighten out the queer effects of this alchemical process of deformation (becoming-werewolf, becoming-studio, becoming-imperceptible as the body takes on the darkness of its milieu). Thus, we can read Le Corbusier's laws as paranoid attempts to support a system of mastery, whiteness, masculinity, and ownership that, despite its best efforts, cannot prevent the clinamen from crashing into its sanitized and controlled laboratory creating a fecund mess.

The Window

If Bandinelli's "Night Class" focuses our attention on the studio's fireplace, Betye Saar's work titled "The View from the Sorcerer's Window" (1966) turns our attention to the window of the studio as another aspect of its dark infrastructure. But before we engage directly with this work, it is important to reflect on Saar's studio as such. Saar describes her studio as a space-time of "play,"[60] in which she experiments with bits and pieces that she has collected over the years. Looking at her studio, it is packed with "ordinary objects," which she "changes the meaning of."[61] This practice starts with extracting objects, such as clocks, from their functional role in society as timekeepers. The whole studio takes on the appearance of a cabinet of curiosities, in which objects are arranged in idiosyncratic clumps that often release form from function. But more importantly, this suspension is also symbolic. For Saar, "an object has its story about its previous life" but through studioing, Saar sometimes "change[s] that," inserting objects into alternative, counter-narratives.[62] The studio renders inoperative functions and meanings, and thus opens up objects to alternative, non-prescribed uses and narratives. The studio is also a space in which time radically alters its linear trajectory. Saar's studio is full of objects from her own past in which ancestors return into the present moment. Saar narrates, "In my studio, I stare at the relics from my family, the faded photos, bits of lace and fabric, gloves, hankies, pressed flowers, and bits of jewelry. The mementos of long-forgotten events, kept for old memoirs."[63] Brian O'Doherty writes that the studio is a special kind of space-time machine in which time becomes a "mobile cluster of tenses, quotas of past embodied in completed works, some abandoned, others waiting for resurrection, at least one in process occupying a nervous present, through which, as James Joyce said, future plunges into past, a future exerting on the present the pressure of unborn ideas."[64] This is not a linear time—the time of operative clocks—instead, it is a non-linear, ouroborotic time in which many temporalities fold in on each other (much more like a spiral than a straight line). O'Doherty adds, "Time is reversed, revised, discarded, used up"[65] in the studio. The strange work of the studio is just as much spatial as it is temporal. Time becomes pliable, plastic, looping back on itself in order to retrieve lingering potentials, waiting in the darkness. Indeed, Saar's studio is a dark ecology of leftovers that have been forced out of historical memory yet remain active, in the shadows. Instead of keeping time, the studio releases time from all measure. The past ceases to be past and instead pole-vaults into the present as if reanimated from the crypt. In sum, the space-time machine of studioing practices is for Saar linked to alchemical rituals of transformation that alter the geometries of space and the trajectories of time. Describing how she moves through the studio, Saar states, "It's more than making an art object, it's like having a reading by a gypsy."[66] Saar's journey in and through the studio animates witchy potentials for sketching out alternative pasts, presents, and futures that are no longer predetermined by imposed, racist narratives. The studio is a place for occult "power-gathering,"[67] a kind of battery, charged for experimentation on the self and its perceptual-cognitive capacities through works that do not abide by laws of functionality so much as by obscure gestures, incantations, visions, and ritualities that rescript who can do what, where, and when.

Works such as "The View" are not simply traces of Saar's occult practices, they actually have the capacity to transform the gallery in which they are placed into experimental studios. In this sense, they refuse to become works of art to be passively contemplated and instead have the capacity to power-gather, interrupting typical looking practices encouraged within the white cube of the gallery. They are extensions of the studio's dark infrastructure, allowing the studio to drift beyond its own space-time coordinates, extending and stretching, as if infinitely plastic. The mere insertion of Saar's window works into these art-world spaces is a kind of research-creation that experiments on and with the space-time of the audiences that encounter it, encouraging ongoing study into the effects of perceptual drift. Indeed, the window in "The View" is not so much a safe view from the outside into Saar's studio as it is a portal through which the viewer is suddenly *inside* the studio looking *out* (or perhaps the very dichotomy of inside and outside no longer makes any sense at all).

"The View" is an assemblage of intaglio etchings placed within a found wooden window frame. Saar, famous for her postwar black nationalist aesthetics, utilizes the window to frame six separate (yet connected) scenes. Starting from the top, we find celestial images of moons and stars against cobalt blue and dark grey nighttime sky. The two central panels depict landscapes that connect the earth to the heavens. These landscapes are infused with alchemical symbols and seem to depict places of occult ritual practices. Finally, the bottom panels offer two hands. The hand to the left is white against a black background, traversed by deep black lines that almost look topographical. More alchemical and esoteric symbols are written on the palm. The hand to the right is black against a white background. It is much more naturalistically rendered, but with an all-seeing eye directly in the center of the palm, gazing back at the audience. The window does not frame a single reality but, as Key Jo Lee might argue in her description of black perceptual drift, a "multidirectional, multimodal, and polydimensional perspectival field"[68] or pataphysical abundance that lies just outside the peripheral vision of white cybernetic capture.

Already with the top two panels of "The View" we get various shades of darkness. There is no simple darkness. Instead, once the eyes are acclimated to low-light vision, blackness becomes full of darknesses. These night shades of darkness are not pitch black, but rather indicate subtle differences within blackness that are missed in Bandinelli's etching. Instead of withdrawing, darknesses *creep up.* Here, each shade is not merely blackness illuminated, but rather the shades give hints at what can be seen when we go deeper into the dark, no longer needing candlelight or fire. Instead, when blackness withdraws, it emits its own kind of dark radiation. The sorcerer's view sees what is impossible to see (colors) in the seen (darkness). In this way, Saar cultivates in the audience a taste for the obscure flavors of darkness that would otherwise lurk just outside of the field of vision.

While there are variants of views on blackness within each panel, there are also variants of scenes. In both cases, Saar indicates that the view of the sorcerer is already a composition of views. *The* view gives way to a series of views that do not stabilize into a unified gestalt but rather unfold into a pataphysical multiverse of overlapping, repetitious forms that are never simply duplicates of one another. The view drifts from its perceptual unity as *a* scene, producing a series of scenes that are now unified not by a subject so much as by the object (the window as apparatus for diffracting reality) through which the scenes are viewed. The window as dark infrastructure holds together a multiplicity that might otherwise drift apart or disperse. As Deleuze and Guattari write, a sorcerer is someone who conjures a pack, a band, a multiplicity.[69] In "The View," the audience glimpses the perceptual experience of multiplicity and how the stability of a single reality gives way to a multiverse, each containing possibilities for fugitive escape (and invention). And these views are not simply subjective hallucinations but are rather amplifications of perceptual variants that are *shared* by the audience, who, in turn, verify the "reality" of this experience. Anne Friedberg charts the history of the window in Western culture from Leon Battista Alberti's famous metaphor of the frame of the painting as an open window (*aperta finestra*) up through the metaphor of the window for the computer screen and thus a slow historic shift from "the singular frame of perspective to the multiplicity of windows within windows, frames within frames, screens within screens."[70] Saar's window is a radical implosion of this entire evolution of the window: It is simultaneously an actual window, a painting with a frame acting as its own metaphorical window, and a multiplicity of windows as if it were a digital screen. The window, the painting, and the digital screen collapse into a single dark infrastructure that is as material as it is immaterial, as actual as it is virtual (and in this sense is truly postdigital).

Through the movement of the panels (so many frames within frames) from top to bottom, we find a visual representation of the alchemical adage "what is above, so below." The landscape acts as a mediation point between the human being and the heavens, the stars, moons. The planets all descend and become miniaturized features of the terrestrial world. The palm of the human hand becomes geographic. The flesh mimetically takes on features of its surrounding world just as the surrounding world becomes imbued with script reflecting the spiritual practices and beliefs of human (secret) societies (a psychogeography).

Furthermore, the studio ceases to be a closed in, hermetically sealed space. The inside of the studio (the hands of the sorcerer looking out the window) are transported outside. In sum, the inside becomes the outside, the sky becomes the earth, the human hand becomes a landscape, and the landscape becomes a human palm. These transgressions are the alchemical work of the studio, which does not abide by the divisions and demarcations that characterize the Western, scientific laboratory. Instead, there is emphasis on mimetic displacements and contaminations that are alchemical and pataphysical impossibilities.

The body drifts away from itself, becoming a depersonalized, anonymous body. The hands, palms facing inward, are displaced from inside the room to the outside of the window. But even stranger still, where one expects a left hand, one finds a reduplication of the right hand. The body is displaced but also duplicated, as if it has budded a second, vestigial right hand. Although doubled, the two hands are far from the same. Instead, the doubling heightens their differences: one black the other white, one a terrestrial map the other holding/ sprouting/channeling an all-seeing eye. Thus, the two hands form a singular plurality, a body that is in two places at once, each hand revealing a hidden dimension that the other conceals. These are, in gloria wilson, Joni Boyd Acuff, and Vanessa López's phrase, "conjuring hands" of curious Women of Color (WoC) engaged in research-creation as an act of summoning the powers of the past in order to create visions of possible futures.[71] For Saar, the hands are curious, meaning searching. But Saar adds another meaning to this phrase. These hands are also curious, meaning weird, uncanny. They pass through the window glass so as to become part of a landscape (as above, so below; as within, so without). They are no longer Saar's hands, but are impersonal appendages let loose from the organization of the body, passing onto the other side of the window, passing from the analogue to the digital screen. Conjuring hands of sorcerers thus bud, divide, displace, and transport—all of which are modalities of bodily drift.

Along with the fireplace, the window is another aspect of the dark infrastructure of the studio, each necessary for creating new geographies that override the policing of boundaries defining when, where, and how bodies can appear. For Saar, the research-creation emerging from the studio is ultimately an embrace of the para/crypto-ontology of blackness as outstripping white cybernetics of surveillance and control. The window is less an aperture for gazing at the outside from the inside (as if from a safe, undisturbed position) and more a portal that opens up the inside to multiple, possible realities (a hidden, esoteric multiverse in six frames). This is a dark ecology of interlaced scales and dimensions, in which what is above drifts into what is below, what is inside drifts outside, what is macrological drifts into the micrological.

But most importantly, the work suggests that the audience, standing in front of it, takes on the position of the sorcerer and that the hands they see in the window become *their* hands. There is thus a strange, depersonalizing drift in which bodies and subject positions lose their rigid distinctions in the darkness that the window emits. The audience becomes part of the medium of the work. Medium in this context has two overlapping meanings. First, there is the medium of the work itself, its material infrastructure composed of wood, metal, and etchings. Standing before the work, the audience is suddenly transported into the alchemist's studio. They enter the frame of the window frame and are thereby invited to take up the position of the sorcerer and the sorcerer's gaze. The work renders inoperative divisions between material medium and its immaterial performance. But Saar, working within a black, American spiritualist context, also seems to conjure up another sense of medium: a spiritual medium, or the sorcerer as medium for contacting dark precursors that lie just outside the perceptual-cognitive apparatus of the human sensorium. The material infrastructure of the artwork thus transforms into a dark infrastructure flowing with occult, witchy powers lurking within its depths. The audience performs the role of the sorcerer, gazing out the window, wondering whose hands these hands have become. Entering into the darkness of the piece causes a drift in ego positions between observing audience and gazing sorcerer, and by doing so, "The View" opens the audience up to alchemical powers of transformation before they consciously know what is happening. By becoming part of the physical medium of the work, they become spiritual mediums. As sorcerers, the audience channels the darkness of the work in the present moment, making it come to life by plugging into and switching on its ritual circuitry. The transcendental darkness of the outside is pulled into the present moment just as the present moment is pulled into the darkness that exceeds it. One can no longer be the disinterested, detached spectator. Instead, one is inside a witchy circle, immersed in a process of becoming-sorcerer, becoming-landscape, becoming-celestial. Spatial drift is therefore accompanied by temporal drift as the audience activates the esoteric rituality of the piece, while the piece activates the audience as part of its medium. Saar reveals how the window, as dark infrastructure, is a space-time machine of escape and fugitive politics into a blackness that renders inoperative the distinctions that support white cybernetic calculation—the very same calculations that differentially construct the bodies of Saar's audience/participants in terms of paranoid, racializing categories and dichotomies that lock them into specific spaces and times so as to police their movements and relations.

By incorporating the audience into the act of studioing, "The View" opens up the hermetically sealed chamber of the alchemist's secret lair and gives its secrets away. These are studious secrets. The studio has turned itself inside out while also opening up the subject position of the sorcerer to anyone who views the work. Thus, the secret inscriptions on the palms of the two hands become studious secrets given away to the public. Their riddles are exposed to a broad, participatory audience who can take them up and experiment with their interpretations. While the knowledge they contain is dark knowledge and thus remains opaque, it is an opacity that is given away as a gift to the audience as initiates in a ritual that no longer is owned by anyone in particular. And while Bandinelli's image of the "Night School" was an exclusive club of white men, Saar's "The View" is an inclusive, public experimentation. This does not mean that everyone will have the same access to the darkness that the work generates.

Nor does it mean that everyone will embrace the transformational potentials of becoming-sorcerer. Different audiences will be subjected to different risks. But the important point is that the rituality of the initiation, which begins with an audience simply looking through the window and thus reenacting the role of the sorcerer is, on a certain level, without qualification—even if this leads to the kinds of anarchic disorganization often associated with misfit alchemists.

Clumping Bandinelli's fireplace and Saar's window together across different temporal, spatial, political, and cultural planes allows us to create a virtual alchemist's studio and to explore the endarkened educational potential of various actions within its space-time. Of course, there is nothing special about the fireplace or the window as both can also be found in the laboratory. Indeed, Le Corbusier would most likely approve of both of these architectural features for helping to banish darkness and shadows. Yet when constellated into the space-time of the alchemical studio through the interplay of rules and drift (as explored in the next gate), the banal and everyday functions of these elements are neutralized, opening up to drifts of dark vitality spilling forth. The point here is that on the surface, the studio might look just like the laboratory (or the art gallery), but with the smallest of drifting swerves, infrastructural powers, desires, and material thus-ness can catalyze encounters with otherworldly precursors always at the very periphery of human perception, germinating in dark ecologies. In this sense, there is nothing unique about the infrastructure of the alchemistical studio. This is not to deny that alchemists utilized specialized equipment in their experiments, such as the alembic. But what we want to emphasize here is that alchemical research-creation does not need special infrastructures to experiment with the gap that separates and conjoins art and science, knowledge and pleasure. Instead, it merely draws out the darkness that is always already lurking, even if it is undetected by Le Corbusier and other lovers of the daylight and its purported transparency, clarity, and enlightened reasonability.

Gate Two

Drift←→Rules, or Defining the Hospitable Conditions for Dark Precursors

Through this gate, we try to determine the experimental nature of studioing. In particular, we argue that this experiment is a combination of two elements—rules and drift—and how they interact with one another forming loops between knowledge and pleasure, head and hand, science and art in the form of inoperative works. Rules, as we will describe them, can be seen as another key feature of the dark infrastructure of the alchemical studio. Indeed, on our account, rules are the essence of such dark infrastructure, much more so than any physical characteristics, as rules (and in particular alchemical recipes and what we will describe as patarules) embody the essentially witchy and occult-like powers of the studio to generate hospitable conditions for the arrival of dark precursors. But rules are only half of the equation. Rules can catalyze drift and sometimes drift can ignite new types of rules. Between the two emerges an ouroborotic notion of the alchemical space-time of experimental research-creation.

This ouroborotic space-time generates experimental possibilities that are distinct from the capitalist sciences, which are dominated by epistemological and sociological experimental models. Epistemological experiments concern what can be known within the parameters of good and common sense. They function to falsify inaccurate theories and thereby contribute to verifiable descriptions of the actual world and its affairs. This notion of experimentation will often dismiss or ignore exceptions that are not statistically significant, or, perhaps more severely, will simply be unable to detect exceptions as they fall outside the rubric of what counts as data. Sociological experiments are less concerned with the outcomes of specific scientific experiments so much as they are concerned with the underlying social, political, and economic conditions that make such experiments possible in the first place (and therefore make them intelligible as experiments). They reveal the hidden networks of power and invisible hierarchies that are operating to maintain certain divisions within sciences and the various discourses of persuasion science uses to legitimize itself. Opposed to these more or less mainstream notions of experimentation, Robert Michell theorizes the "ontogenetic" experiment. Inspired by authors such as Bruno Latour and Donna Haraway, Mitchell describes this third kind of experiment as generative of "new entities and assemblages" that are brought into being.[1] Mitchell highlights how Romantic poets exemplified ontogenetic experimentation in terms of "artistic experiments"[2] that exist at the point of contact between "scientific" experiments and artistic invention. For us, another word for ontogenetic experiment is research-creation, or an experiment that concerns itself with rules for producing new ways of sensing the world (new flavors of becoming) and inoperative works (that do not play by the laws defining the ordering of things, insisting instead on their potentiality for otherwise becomings). Below we will focus on how the circulation of rules and drift generate the curved space-time of alchemical and pataphysical studioing, whose composition is particularly conductive to visitations from dark precursors. In conclusion, we offer a brief reflection on a "research-creation happening," which we hosted at our university, that reconstructed the classroom as a collective studio for experimenting with the oscillation between drift and rules. This experiment was an attempt to suspend and render inoperative the dominant logic of learning (and is algorithmic rules of assessment) in the university in order to institute another mode of educational life that, as we will describe below, offered a glimpse into the potentialities of studious drift to displace bodies, thoughts, and gestures in such a way as to open them up for new, unexpected use.

Rules

For Lorraine Daston, rules have largely been split into two camps that, in turn, take on a certain hierarchy of importance in contemporary life. On the one side, we have rules-as-paradigms. Rules-as-paradigms are thick, meaning that they are accompanied with many examples and exceptions and are distinctly related to context. Daston points out that these rules differ greatly from contemporary rules insofar as they "anticipated their own incompleteness"[3] and as such incorporated into themselves exceptions. Paradigmatic rules, in pre- and early-modern Western society, were attached to specific trades and arts, and developed over time in relation to know-how, tact, and sensitivity toward contextually specific variables. Indeed, exceptions *were the rule*. Also, these rules did not operate in terms of induction or deduction so much as by analogous reasoning whereby similarities across different, singular cases were of central importance (hence the lists of examples often accompanying these rules). Instead of rigid and strict, these rules were pliable, soft, and flexible, demanding a certain amount of discretion before being applied to a particular case. The post-classical Latin meaning of *discretio* included "prudence, circumspection, and discernment in weighing matters."[4] To be discerning meant one had to cultivate the taste for analogies (in which particulars could be grouped in loose constellations according to various family resemblances), particular situations (in which rules had to be modified in order to be deployed), and exceptions (in which rules had to be suspended). Following a rule was thus an act of "imitation-without-copying."[5] Each case demanded its own modifications and variations. Moving somewhat beyond Daston, we might say that rules-as-paradigms concerned intensities, or unmeasurable qualities of a feel for "more or less."

On the other side, Daston places rules-as-algorithms. With the rise of colonization, urbanization, and the capitalist division of labor, the rule-as-algorithm became increasingly dominant, displacing (and ultimately marginalizing) the rule-as-paradigm. Indeed, Daston makes the strong claim that we now live in an "empire of algorithms."[6] The rule-as-algorithm is thin, meaning it lacks examples and exceptions. Instead, it is a maximally generalizable and abstract formula that can function in a context-free environment. It therefore does not necessitate sensitive discrimination, and in fact, automates its own functioning so as to erase the need for subjective interpretation and judgment. In this sense, we can connect the empire of algorithms with Daston and Galison's notion of mechanical objectivity, which we discussed in the initiation to this book.[7] Both separate head from hand to the point where even the head seems to disappear into an increasingly detached mode of calculative reasoning. Although Daston does not argue this, we might even go so far as to say that the rule-as-algorithm also erases the object, replacing it with a calculation (not unlike the power of exchange-value over and against use-value in a commodity). Rules-as-algorithms focus on quantifying intensities (immeasurable excesses), or transforming intensities into extensities (measurable amounts). The true violence of the rule-as-algorithm is not found in Daston's work but rather in the analysis of logistics provided by Harney and Moten.[8] For them, the brutal economy of the algorithm underlying the modern logistics of shipping has its genealogical origins in the hold of the slave ship.

The dark infrastructure of the rule in alchemical studioing is somehow located in the paradoxical space between rule-as-paradigm and rule-as-algorithm, or rather in the exceptional point of contact in which the division ceases to make sense. Take for instance rules in books of secrets, and in particular the abundant alchemical recipes contained within. These rules demanded a lot of discretion on the part of the alchemist who had to interpret a gamut of arcane references, solve riddles, induce various trance-like states, and, on a more practical level, simply determine actual references to lists of materials that were often cloaked in alchemical *Decknamen* or cover names. Did "lead" refer to actual lead or to some other substance called "lead"? Likewise, the number of ingredients in an alchemical recipe might vary. Did "one" ingredient refer to "one substance" or to "one class of substances"? In all cases, as Principe writes, these variables in recipes introduced a certain amount of textual glissade wherein "'right' meaning constantly slips out from under the feet of [the] interpreters."[9] Discretion was called for in the handling of such texts, their interpretations, and in the relation of texts to material experimentation. At the same time, as demonstrated in William R. Newman and Lawrence N. Principe's detailed study of seventeenth-century notebooks of George Starkey (the American alchemist), recipes also encoded consistent and systematic methodological practices such as exacting observation, data recording, and data analysis (we might say algorithmic-like rules) concealed behind layers of distraction.[10]

In this sense, alchemical rules did not abide by either rules-as-paradigms or rules-as-algorithms. It would therefore be a mistake to simply see the various symbols and riddles used as mere distractions that can be peeled away in order to reveal the scientific "truth" hidden behind occult smoke and mirrors. Instead of finding the value of alchemical rules in their nascent scientific method and rational investigation into metals, we want to value the occult features of these rules that paradoxically bring together visionary states/dreams/

symbols and experimentalist rigor (without statistical measure) in order to produce inoperative works that challenge divisions between art and science, head and hand, pleasure and knowledge. Thus, we take some umbrage when Newman and Principe summarize their scientific analysis of the experiments of alchemist Joan Baptista Van Helmont as follows: "What he [Van Helmont] saw in his furnace was not a manifestation of the development of his own soul or a vision of his unfolding psyche,"[11] which Van Helmont recorded in his book *Ortus*. While Newman and Principe decode Van Helmont revealing objective chemical insight, what is lost is how Van Helmont's practice also and equally was a work on the self, on what can be seen, on how the invisible can be made visible through visionary practices, how rules can transform the cognitive-perceptual apparatus of the alchemist, precisely through rigorous experimentation with the furnace. From Van Helmont's perspective, which is dismissed by the "enlightened" authors, there is no division between visionary arts and scientific investigation. The inoperative (meaning nonfunctional, vestigial) organs that Van Helmont cultivated through the rules—and which do not fit nicely into Newman and Principe's narrative of scientific exactitude and objectivity triumphing over occult superstition and hermeneutical obscurities—offer up sensations and visions that are unintelligible to those bound to algorithmic rules, but for this reason, they are all-the-more precious. Alchemy has been dismissed and neglected precisely because of the occult features that, on our readings, offer radical potentials for rethinking the very nature of rules that now structure the science and art divide.

We want to hold true to the meaning of "recipe" as a way of writing out rules for experimentation. Recipe rules cannot operate as principles but as models to be emulated, meaning that they have to be creatively tinkered with, discriminately put into practice, and used in order to see what they make possible.

Pataphysical authors, such as Alfred Jarry, are perhaps the most important inheritors of the generative role of rules within alchemical traditions. As Katie Price points out, in pataphysics there is a unique investment in systematic and exhaustive rule experimentation "grounded in jocoserious [parodic] constraints"[12] that "fuse rationality and irrationality, procedure and chance, humor and seriousness."[13] For Jarry, the pataphysical experimentation concerns "the science of the particular" (instead of the general) and attempts to uncover and elucidate "laws governing exceptions."[14] In the case of the former, pataphysics bucks up against modern science, which concerns generalizable laws, and in the case of the latter, it attempts to theorize a law regarding that which would otherwise be lawless (the exception). This means that pataphysics can only progress analogically (proceeding from particular case to particular case) rather than inductively or deductively. Thirdly, pataphysics can only offer "imaginary solutions"[15] that do not abide by the logic of good or common sense and often produce more problems/questions than answers and/or posit solutions to questions that have yet to be formulated.

In pataphysical sciences, exceptions are not excluded from the rule but are the rule. Indeed, experiments are meant to induce exceptional states and to produce exceptional objects that have no equivalent within the order of things (as Foucault would say). As Christian Bök states, pataphysics does not study rules governing general recurrences of *"expected* cases" so much as they induce *"excepted* cases."[16] In particular, they are interested in three kinds of exceptions: variance, alliance, and deviance. Variance concerns anomalous cases of incidents that are surprising and disruptive to expectations. Alliance includes *syzygia* or *discordia concors* between seemingly unrelated concepts, words, gestures, images, bodies, and so forth. And deviance concerns the awkward drift of the clinamen, producing detours or disorientations that fall offline or off track.

Pataphysicians formalize constraints in order to swerve or warp time and space, opening up these dimensions of experience to pataphysical trapdoors and escape hatches. The constraint is another name for what we prefer to call a *patarule* or a rule pushed to an absurd dimension so that the rule that rules the rule reveals itself, ultimately becoming something of a parody, not in the sense of a "joke" but in the sense of a rule that no longer can calculate its own effects, that no longer has strict control over itself. Patarules suspend or render inoperative calculative functionality in order to expose the rule of all rules (that is, algorithmic reliability, efficiency, and predictability) to an *internal*, aleatoric swerve that cannot be predicted in advance.[17] The patarule is thus an algorithmic rule *as not* an algorithmic rule (or, as a later gate of this book will explore in more detail, the algorithmic rule becomes postdigital even as it sustains the formal qualities of calculative reason). Alternatively, we can think of the patarule as embracing the darkness below the apparent readability, efficiency, and predictability of the algorithm.

For instance, Jarry produces a series of constraints for how to build a time machine that disregard conventional divisions between math and poetry. These patarules appear to obey strict scientific principles and thus appear at first blush to fully embody algorithmic exactitude, freed from contingencies of language, context, and specific interpretive use. And yet, the rules do not work. They do not provide accurate and reproducible calculations that would enable the time machine to actually function. In this sense, the exactitude of the rule-as-algorithm turns out to be a parody of itself. This algorithmic form is not negated but rather preserved in a suspended or inoperatively vestigial form. And by doing so, the rule-as-algorithm flips back into its proverbial other: rule-as-paradigm, which concerns itself with exceptions and necessitates discretion and hermeneutical interpretation. Pataphysics is the study of exceptions to laws, but it does so through rigorous deployment of algorithmic rules drawing from math and physics. Yet, these algorithmic rules do not function as one would expect. Instead of generating predictable outcomes, they spit out singularities that resist becoming subsumed under any general type, species, or genus. They are monstrous excretions born from an algorithm that has lost its functionality. Thus, the scientific function of the rule-as-algorithm suddenly (alchemically) transforms itself into the aesthetic rule-as-paradigm. And for this reason, the rule-as-algorithm needs a certain amount of poetic discretion in order to be understood and made pleasurable. This is not an irrational use of algorithmic rules as much as it is "surrational,"[18] which, for Bök, involves *rigorous* study of science and mathematics in order to *precisely* induce a swerve or clinamen where it is most disruptive and productive and therefore makes the most difference with the smallest amount of force. In sum, Bök writes, "the pataphysician does not counteract science so much as exaggerate science, adopting it parodically and applying it excessively . . ."[19] through experimentation with constraints such as those described by Jarry in his time machine blueprints.

The difference between the rule-as-algorithm of science and its exaggerated parody might be a minimal difference that nevertheless makes all the difference (a difference that is infrathin, as Duchamp might day). The swerve that Jarry induces renders inoperative algorithmic functions from within the function itself, producing a weird pleasure where knowledge should be. This means that knowledge (tested and verified through mathematical modeling and algorithmic calculation) without pleasure transforms into non-knowledge of a pleasure that does not abide by the traditional categories of aesthetic taste (such as the beautiful). Dark, (an)aesthetic pleasures accompany dark knowledge (or non-knowledge) of inoperative things that offer "impossible solutions" that, in the end, produce more questions than answers. Jarry's time machine does not signal progress in scientific knowledge, nor does it offer a practical intervention into transportation. It also does not offer expected and familiar aesthetic pleasures that are easily recognized in relation to standards of taste. Instead, it (re)animates alchemical experiments in producing impossible objects through techniques that prefer not to operate with the assumption that science and art, knowledge and pleasure are separate (and unequal). The result is an anomalous, singular exception that, from within the current empire of algorithms, is useless, impotent, unproductive, and nothing more than a misappropriation of mathematics and/or a "failed" aesthetic experiment that simply leaves a bad taste in one's mouth. But what we find important is how the time machine is exemplary of how discrete spin or swerve can produce unrecognizable yet palpable effects in partitioning of the sensible, the order of things and divisions of labor, prompting the invention of new languages, new tastes, new flavors, and new forms of (an)aesthetic (non)knowledge.

What Jarry has thus provided in his blueprint for making and using a time machine is actually a *blueprint for a studio* as a special kind of space-time machine that is untimely dark and atopic (not unlike Dr. Who's Tardis). He describes the ability of the time machine to "isolate us from Duration, or from the action of Duration (from growing older or younger), the physical drag which a succession of motions exerts on an inert body."[20] The studio offers a slight displacement of space and time so that the duration of the world can be studied. The studio is a d-rift. It is adrift in the sense that it does not appear where it ought to appear or when it ought to appear. It therefore crosses boundaries between compartmentalized spaces and times, largely unaffected by the world's duration. And it is a rift in the underlying (algorithmic) rules that bind space and time to particular bodies, particular ways of knowing and sensing, particular distributions of gestures and actions. The studio itself is therefore an *impossible* solution to the question of the relationship between science and art, pleasure and knowledge. It is impossible in the sense that it does not make sense within an empire of algorithms (as it cannot be predicted when and where the studio-as-space-time-machine will open up and unfold). And yet, it persists as a crypto-genealogical inheritance as a special kind of teleportation infrastructure capable of inducing alterations in space and time that invite dark precursors (from ethernity and thus not constrained by worldly duration). In this way, patarules are rules of engagement with the darkness of the clinamen.

This studio space-time is distinct from the space-time operations of the laboratory. In the classic sociology of laboratory life offered by Bruno Latour and Steve Woolgar, the laboratory is described as an enclosed space-time apparatus for ordering disorder in order to construct facts. Rules-as-algorithms are an essential way in which scientists struggle to produce order out of a disordered array of observations. Laboratories produce a variety of statements, and at the beginning of a research process, these statements rely upon the discretion of the observer. Rules-as-paradigms must take into account the judgment of the author of the statements. Yet, through various procedures, science attempts to reduce such discretion, and thereby erase the author of statements in order to produce facts that speak for themselves. In this sense, rules-as-paradigms are slowly replaced by rules-as-algorithms, which leave no trace of authorship. Rules-as-algorithms are contextless, authorless, and refer to something "out there."[21] Through this process, facts become reified and take on the appearance of something beyond sociological or historical analysis and thus free from temporal and spatial contexts of production. Rules-as-algorithms slowly replace situated judgments of the discretionary type with objective facts of the matter through the deployment of hypotheses testing, proof, and deduction so that observation and inscription correspond with formal logic (which is non-perspectival and value free). One set of rules operates within and simultaneously produces *places* with specific features while the other operates within and simultaneously produces abstract *spaces* that are homogenous and featureless, endlessly replicable because of its simplified, geometric purity. The first is the space-time of decisions while the second is the space-time of facts. Summarizing the outcome of this process, Latour and Woolgar write, "The result of the *construction* of a fact is that it appears unconstructed by anyone; the result of rhetorical *persuasion* in the agnostic field is that participants are convinced that they have not been convinced . . . as to the *circumstances*, they simply vanish from accounts, being better left to political analysis than to an appreciation of the hard and solid world of facts!"[22] We can then contrast this description of the work of the laboratory with the work found in the studio. For us, the studio is a permeable (rather than enclosed) space-time for disordering order (rather than ordering disorder) for the purpose of inviting in the swerve of a dark precursor capable of generating exceptions. While the goal of the laboratory is to construct "solid" facts that are useful, the studio suspends this goal by continually producing exceptions and expenditures that are useless and impossible. Weirdly enough, in the final chapter of their book, Latour and Woolgar offer up an image of an *inoperative* laboratory that we want to highlight as a kind of pataphysical, parodic studio. They imagine a scenario in which the enclosed, purified, ordered, algorithmically bound laboratory is invaded by an outsider who peels off all the coded numbers from samples and returns them to a refrigerator, unmarked. "This nightmarish experiment," write Latour and Woolgar, "highlights the importance of the trapping system" that fundamentally defines the laboratory.[23] The laboratory must produce order out of disorder (quantity out of quality). Thus, a deviation in which entropy is increased through a swerve of a dark precursor can only result in "extreme confusion"[24] causing the whole laboratory to d-rift from itself becoming something else, something imperceptible and unintelligible to science.[25]

In short, both alchemy and pataphysics live in the awkward yet generative space and time between rules-as-paradigms and rules-as-algorithms, refusing to substitute the latter for the former. The historical relationship between the two can be figured as such: In alchemy, we discover the emergence of the rule-as-algorithm within the rule-as-paradigm (the recipe), and with pataphysics we discover the sudden appearance of the rule-as-paradigm within the rule-as-algorithm (the patarule or constraint). In both cases, this oscillation between types of rules means that questions of pleasure cannot be separated from questions of knowledge, and questions of art/creation cannot be separated from questions of science/discovery. As opposed to Latour and Woolgar's analysis of the laboratory in which rule-as-paradigm is *replaced* by the rule-as-algorithm as the gold standard for what counts as a fact, here we have a series of ouroborotic swerves or entanglements in which the two are never fully separated but rather deactivate/activate one another in complex (and ultimately impossible) points of contact. The dark infrastructure of the studio (rules, windows, fireplaces, and so forth) incubates awkward drifts between these two. What the laboratory attempts to stabilize through the empire of algorithms, the studio intentionally sets adrift.

Rules Adrift, Drifting Rules

At this point, it is imperative to turn to the exact relation between rules and drift and how they enter into a certain feedback loop that knots together into the form of an impractical rituality—a rituality that does not eliminate the clinamen but rather keeps open a space and time in which a dark precursor is invited to enter. In other words, we want to take time to develop the ouroborotic movements of ontogenetic research-creation in the space-time of the alchemical/pataphysical studio. This is not a story that can be narrated in any linear fashion with a beginning, middle, and end. Instead, one must think of the ouroboros who is eating its own tail: Rules of engagement induce drift, and drift summons from ethernity new rules for emergent forms-of-life. Which came first, the chicken or the egg? Ouroboros. To illustrate this point, we will turn to three different relationships between drift and rules found in contemporary philosophy. Each, in its own way, captures something of the back-and-forth oscillation between rules and drift. After introducing the three frameworks, we will then propose our own, synthetic philosophy that attempts to open up a new dimension in our understanding of drift in terms of its ontology and intensity.

Drift of Desire

First, Deleuze and Guattari define the schizophrenic in terms of the drift of desire. The schizo "explodes the Oedipal genealogy" that forces desire into the limiting confines of a bourgeois family drama and instead provides a "generalized drift" in which nothing is original or derived.[26] Within this drift of desire, "everything comingles" in "intense becomings, passages, migrations"[27] that break through dichotomies between self and other, inside and outside, before and after, individual and collective. While it is popular to think of Deleuze and Guattari in terms of deterritorialization of desire as a productive force and thus antithetical to any notion of rule (as a territorialization), Deleuze and Guattari are clear that the schizophrenic deterritorialization of Oedipalized desire is similar to what Claude Levi-Strauss defines as *bricolage* which is organized around "rules of thumb which are fairly extensive, though more or less a hodgepodge— multiple and at the same time limited; the ability to rearrange fragments continually in new and different patterns or configurations; and as a consequence, an indifference toward the act of producing and toward the product...."[28] The rule here is a flexible and open-ended rule of thumb rather than an absolute, sovereign law. It is a rule that is extensive, multiple, but not necessarily generalizable (it is "limited"). Indeed, the rule of thumb is immanent to a set of particulars. It is also a hodgepodge and therefore internally inconsistent if not contradictory, demanding continual adjustment and decision/discretion. And it is productive of combinatorials not unlike an algorithmic formula, even if it is ultimately disinterested in operativity. Instead of being interested in producing (functionality) or the product (a measurable, predictable output), the rule of thumb is in a conspiracy with the experimental vector of desire's drift. This is not a rule from above for managing schizophrenic neurodiversities so much as a rule from below for experimenting with what neurodiverse states of being can do. Thus, rules of thumb are necessarily internal to the drift of desire rather than merely external impositions.

Perhaps we can characterize such schizo-drift as becoming-metallurgical of the subject. Deleuze and Guattari describe the art/science of metallurgy as the discovery of the "flow of matter."[29] In the forge, bodies undergo processes of "continuous variation"[30] across a series of states of intensity (between hot and cold, hard and soft, ridged and mailable). Such continuous variation is not exclusive to metals but rather exemplifies a "life proper to matter, a vital state of matter."[31] Metallurgy becomes the consciousness of "matter-flow"[32] or what we might call matter-drift that defines an expansive terrain for all nonorganic life. Metal, in this sense, is not an object or thing so much as a body without organs, or a body that is composed of intensities and flows rather than compartmentalized, hierarchically structured parts/organs (each with its own function in relation to other functions). Here, Deleuze and Guattari return to alchemy. They write, "The relation between metallurgy and alchemy reposes not, as Jung believed, on the symbolic value of metal and its correspondence with an organic soul, but on the immanent power of corporeality in all matter, and on the esprit de corps accompanying it."[33] The fundamental problem with Jung's psychoanalytic interpretation of alchemy is that it *Oedipalizes* matter-flow, trapping it within a series of psychological myths and symbols. He fundamentally did not recognize how matter-flow escapes from such symbolization and therefore is not simply expressive of various states of the soul but is also a productive power of inorganic life to produce variants through drift. Alchemy is thus reclaimed as a precursor to schizoanalysis, and the drift of desire becomes simply another variant of a more basic material drift. Indeed, desire is a becoming-metallurgical of a subject (as a body with organs), and this is accomplished in/accompanied by the space-time of the alchemical studio, which unleashes such drift through experimental rules that apply to the metallic dimensions that pass between and through subject and object. Because alchemy deals with accidental forms (intensified, metallurgical drift), it can never function according to algorithmic rules alone. Instead, it demands rules-as-paradigms capable of dealing with situations that are *more or less*. The alchemical space of the more or less is a nomadic space defined by desire's lines of drift.

The Useless Drift of Life

For our second model, it is important to point to the ongoing influence of Fernand Deligny, a pioneer of the anti-psychiatry movement in France. Through his nomadic educational experiments outside the space-time of the psychiatric clinic, Deligny worked with autistic children, but he never attempted to normalize them or to write their biographies. Instead, he recorded the routes of their embodied movements in what he called "lines of drift."[34] The intricate and sensitive drawings of the wanderings of autistic children—sometimes overlaying multiple routes of a single child and sometimes overlapping routes from multiple children to form a composite image—were influential to Deleuze and Guattari, who saw in these drawings the complex interweaving of traces within maps, and thus the origins of their own theory of nomadic space.[35] When superimposed on top of one another, the various circles, swerves, and curves appeared to loop into a consistent tangle of routes that, for Giorgio Agamben, reveal the shape of a life or how a life uses itself to define itself. A life adrift is a life in use. Use in this sense is an "inoperative praxis," which neutralizes any work, occupation, or action that would come to define what counts as a life before the life puts itself to use (lets itself drift). The drift is not random or chaotic, it is patterned in a certain way that is found useful to the individual for defining whomever and whatever they are. As such, it is not as though drift negates all rules for how a life might compose itself. Instead, the inoperative praxis of drift is precisely an "indetermination between rule and life"[36] so that life takes on a certain form, and a certain form is brought to life (through whatever shape its drift might take via its routes). Life is defined by and through how it uses itself, how it loops into itself and drifts away from itself simultaneously. A life in use is useless to institutions that attempt to normalize (and make bodies march to a specific set of predetermined, neurotypical patterns/laws of socially acceptable behavior). Thus, there is something impractical in the practice of a life in and through drift. Such a life is not defined in terms of ends. Instead, it is a means released from ends so that it can experiment with what becomes within the drift. Deligny's radical resistance to institutionalizing autistic children and his insistence on mapping their lines of drift was, in this sense, an effective way to demonstrate the fundamental drift of studioing, which is always in excess of all institutional forms, practices, and sanctioned/legitimated actors.

Erin Manning also connects drift to uselessness, and uselessness to neurodiversity (including the autistic children that inspired Deligny's drift line drawings). Writing about one of many experiments in her SenseLab called nestingpatching pocket practice, Manning highlights how its outcomes are "absolutely useless."[37] Uselessness in this context means that nestingpatching pocket practices produce studioing space-time drifts in the typical classroom, "claiming not a space of intelligibility but a thresholding into what is not yet known."[38] The experiment in warping space-time "claims the useless"[39] drift of ideas, gestures, and materialities that are "often bulldozed by the frontality of the interpersonal and its ways of guarding the neurotypical norm and the whiteness it celebrations"[40] through the spatial-temporal enclosures of the typical classroom. For Manning, the studioing of nestingpatching pocket practices is an embodiment of neurodiverse swerves in the learning apparatuses of the university classroom, which reinforce neurotypical whiteness as a default. It is important here to note that Agamben and Manning seem to be arguing that neurodiverse routes/patchworks are *paradigms* for life in drift. As exceptions to the normative rule, neurodiverse materialities exemplify or make intelligible what might otherwise remain hidden in average-everyday life: that life as such drifts, which demands continual experimentation on and through the use of rules to produce the routes that compose a form-of-life.

Drifting Inflections

And finally, Manning argues that drift is found in the infrathin distinction between intention and inflection.[41] An intention is volitional whereas an inflection is an unexpected change of direction on a curve activated by an intensity of invisible and imperceptible forces. Through this change in curvature (drift), the subject is moved beyond themselves (as defined by their intentions). Drift is the minimal space and time to pry open the subject to be able to receive the swerve effects of various intensities (dark precursors) on their habituated (neurotypical) movements and gestures. While this might seem antithetical to rules and rule-making, Manning points out that there is an element of planning when it comes to inviting the dark precursor to pay a visit. She writes, "Planning in the undercommons is the setting into place of techniques that open the event to its more-than, the setting into place of conditions through which the minor gesture expresses itself, tuning the event to its potential."[42] When discussing the activism of the indigenous group Idle No More, Manning clarifies this relationship between rules and drift. She observes, ". . . rules are invented in the event of the game" and "this we learn from our children whose games morph in the playing, and for whom the art of fabulating is still very much alive."[43] Fabulation is "setting into place" conditions for curving. For us, this is the space-time machine of the studio. Such fabulation is not arbitrary but is instead a making ready through certain rules. These rules are not transcendental laws that are unaffected by what happens (and thus somehow outside the situation at hand). Instead, they are immanent to the field that they shape (and are shaped by). As such, there is a minor difference between the rules of the game and the playing of the game.

Again, Manning emphasizes the connections between drift, inflection, and neurodiversity. For her, "artfulness" is a sensitivity or openness to newness, and autistic perception is "the direct participation in the event, of its welling ecologies, is perhaps the most open register for the experience of the artful."[44] Autistic perception is particularly artful insofar as it is attuned to resonances that would otherwise fall under the radar of the perceptual-cognitive wiring of neurotypical individuals. Autistic perception (in the critical sense) is dark perception, alchemical perception of perceptual drifts that arise through inflections (rather than through intentions).

In all three instances, we see an intimate relationship between drift and rules. Drift takes on a form (through experimentation with rules), and rules produce more drift in what a subject can do, perceive, think, and make. As such, there is a loop that begins to take shape, a set of entanglements that both constitute a turning (Agamben's routes) while also maintaining a certain capacity for shifting the parameters of what is knowable and enjoyable within a situation, producing neurodiverse *pata-psyches* capable of perceptual attunement to the inoperative dimension of objects, gestures, and languages that are brought into the space-time of the alchemical studio.

Rules and the Spatio-Temporality of Drift

We are inspired by the authors we have briefly covered, but we would also like to add to this discussion two small but important amendments concerning the ontology of drift. First is a spatial clarification of drift that is absent in Deleuze and Guattari's use of the concept, and second, we want to highlight the paradoxical temporality of drift. In the first case, rules interact with drifting flows in order to amass loose, porous clumps (inoperative works) and/or rules emerge from within these clumps to set forth additional drifting movements, and in the second case, rules manage the speed and intensity of drift so as to avoid radical acceleration (and thus the complete annihilation of all works). Again, what is at stake in this analysis is the contact between rules and drift that constitute the space-time of alchemical and pataphysical studioing.

The Space of Drift: The Clump

Let us start our analysis of the spatiality of drift by returning to Deleuze and Guattari. In the original French publication of *Thousand Plateaus*, Deleuze and Guattari utilize concepts such as *flux* and *dérive* (with *flux* being the much more common concept), which are often translated as flow and drift, respectively. *Flux* is an interesting concept in that it speaks to a more algorithmic or quantitative understanding of movement than *dérive*. It has distinct connections to thermodynamics, for instance, and to questions of scientific measurement. Although Deleuze and Guattari do not address this issue explicitly, it is possible to think of *flux* as in constant tension with rules-as-algorithms: both that which is captured and codified by algorithms and what attempts to escape them. On the other hand, *dérive* is much more qualitative, situated, and lends itself to the idea of movement as flow. Indeed, *dérive* originally meant to change the natural flow of a river (*de* = away/*rivus* = river). This concept offers a situated understanding of water and its drifting path as related to contextual specificities (think of a farmer who is working intuitively through their familiarity with the land to redirect a flow of water). If this is the case, then flow is kind of movement that is in constant relation to rules-as-paradigms, which concern measurement, calculation, and exactitude. *Dérive* emphasizes that drift is always "away" or beyond capture by algorithms. If flux is a way, then drift is away from this way, it is a curve (even if this curve is almost imperceptible). Thus, the two terms utilized by Deleuze and Guattari are not necessarily interchangeable but speak to two different ontological flows of becoming that call forth, butt up against, and escape from different kinds of rules.

But what we are most concerned with is that neither of the French terms actually capture the internal dynamic of the English word *drift*, which is both a verb *and* a noun. Importantly, the etymology of drift combines the old Norse word meaning a mass of snow or leaves and the Middle Dutch word for course. Drift is thus simultaneously a slow, haphazard clumping together of tumbling, wayward elements *and* movement. It is a word situated between a place where things gather in common *and* a time of dispersal. Drift is a situated push and pull between holding onto and letting go. The clump is a dynamic heaping, a composing of mobile bits and pieces without firm boundaries or borders. Its edges are frayed. The clump is a temporary organization that is always already moving away from itself, drawing itself out to the point of dispersion. It is a network of analogically relates bits and pieces that temporarily come to rest, and yet continue to vibrate with an untapped potentiality for more movement. Think of the drift of snow and the snow drift: the movement of drift generates piles that cumulate here and there in an *ad hoc* fashion. These clumps are loose structures that are never clearly defined. They are porous masses that are continually open to more drift in and out of the clump. This is not a dialectical relationship as it does not indicate a negation of one state by another, but rather indicates a continual, ouroborotic oscillation. As Lauren Berlant once said, "Objects are always looser than they appear" and "objectness is only a semblance, a seeming, a projection effect of interest in a thing we are trying to stabilize."[45] Objects are drifts, loose clumpings that only seem to stabilize for a moment before they drift away from themselves.

In a way, we might say that employing "drift" to translate the French is something of a fortuitous accident insofar as the English term brings with it two meanings simultaneously. Deleuze and Guattari, we feel, would appreciate this slippage of meaning as it seems they were constantly attempting to understand how territories deterritorialize and reterritorialize, which is not unlike the internal relations eloquently summarized in the single, more intuitive sense of drift.[46] In a similar manner, for Agamben, routes tend to heap up and these heaps of life become the frayed and fragile form of *a life* (a porous clump of recursive lines and rules looped together). And looking back over her collective, nestingpatching pocket practices, Manning reflected, "After a few years, in a collective state of confusion, having at best been able to anarchive the process of anarchiving, we succumbed to our own messy archive of images, videos, recordings, writings, and artworks amassed over a decade of experimentation on four continents to ask whether the archive itself might be a departure for what else moves through it."[47] For her, the anarchive is the place where the residue of archiving forms a clump or a drift of materials. The anarchive as a drift has a temporary solidity/form yet ultimately drifts away from itself and thus cannot be configured as hermetically sealed, carefully curated, and perfectly preserved repository for "Truth." Furthermore, unlike the archive, which is an intentional, highly structured and cataloged form, the anarchive as a drift is largely an inflection point of a process and thus the outcome of an internal dynamic that might not have been intended at all. In all the authors covered, two meaning of drift are implicitly implied in their analyses. For us, this is an important point to emphasize because it further specifies the contact point between rules and drift: Rules can produce clumps out of drift (intensified territorialities of desire, forms-of-life, anarchives) or they can set adrift movement out of temporarily stabilized yet porous clumps.

Returning to Moten and Harney, the hold of logistics over the drift of our desire is impossible precisely because digitized objects are themselves loose. Even in the slave ship, "imagination produces nothing but exsense in the hold" where ex-sense is a surplus sense that introduces a swerve in the algorithmic logic of dehumanizing logistics, returning us to the haptic intimacies of discretionary rules over that which drifts. The hold—in which the empire of algorithms holds human life captive—also implies a haptic hold of an emergent spatio-temporal undercommons. For us, the two meanings of hold found in Moten and Harney's work gestures toward the spatial and temporal dynamics of drift: the hold of logistical holdings is never complete but rather contains within it a multitude of "outlawed, interdicted, intimate things of the hold, containerized contagion," in which "logistics externalizes logic itself to reach you, but this is not enough to get at the social logics, the social poesis, running through logisticality."[48]

The Time of Drift: Slow Haste

Equally important is the unique temporality of drift (both as it clumps and as it moves) in relation to rules. For some who read Deleuze and Guattari as simply encouraging "free flows" of desire that are radically deterritorialized, the speed of such flows can quickly produce an ideology that uncritically praises pure accelerationism (more on this topic in a later gate). But this accelerationist reading of Deleuze and Guattari does not take into account the dynamic interplay between drift and rules, which we have been outlining, nor does it consider how drift (unlike flow) emphasizes temporary accumulations rather than unimpeded accelerationism. Clumps, in other words, are accidental resting spots, or even better, points at which movement slows to such a degree that it becomes imperceptible without necessarily stopping. In the space-time of the alchemical/pataphysical studio, drift-rules-clumps form a circuit that has its own s-pacing (space + pace). To understand this s-pacing, it is helpful to compare and contrast it with modern and postmodern understandings of movement as outlined by Byung-Chul Han. Modernity, for Han, emphasizes "marching toward a goal" as the primary movement of modernity. This type of movement is guided by a (more or less linear) "teleology of progress," which produces a certain pressure to "accelerate" in order to achieve one's goals.[49] Here we can think of the linear story that scientists tell themselves about the work of the laboratory, which is progressively attempting to overcome disorder with order, individual, context-sensitive discretion with impersonal, context-free algorithms. The march of reason is directed toward the goal of constructing facts that correspond with reality "out there." Or one can think of the dialectic of enlightenment, in which such marching orders transform into the terror of fascist political marching.

But in postmodernity, we are experiencing a shift from marching to *whizzing*. Whizzing, for Han, is the form of movement that emerges when there is no telos to direct us toward a goal and against which we can measure progress (or regress). Whizzing is phenomenologically characterized by "haste and restlessness."[50] Whereas the former dictated an imposed orientation, the latter lacks any orientation whatsoever, normalizing a feeling of disoriented irritation or free-floating waywardness. The result is an experience of being perpetually rushed without being able to grasp where one is going. Whereas the march is the space-time of modernity's purposeful, linear movement, Han pinpoints the Internet as the space-time of postmodern whizzing. He writes, the Internet "is woven from possible connections, or links, which do not fundamentally differ from each other. No direction or option has an absolute priority over the others. Ideally, a change to direction is possible any time. There is no finality. Everything is key in limbo."[51] There is no temporal duration in this space, only a series of disconnected nows. And finally, Han argues that on the Internet there is "no work, no completion, but only continually new versions and variations."[52]

Opposed to these two forms of movement, Han suggests his own notion of "contemplative lingering" that emphasizes duration (over and above the brisk pace of production and consumption that marks the Internet).[53] Han summarizes, "The *vita contemplative* is a practice of duration. It founds an *other time* by interrupting the *time of work*."[54] Such a life is a "friendly life" that makes place for something to take place, for something to arrive and stay, for a spatial sense of place to emerge from within a flow of movement.

While we have much sympathy for Han's analysis, we also find a certain romantic nostalgia for a pre-technological, natural world, somewhat disturbing. For instance, one of his key examples of the temporality of lingering is Heidegger's country path, which "illustrates a *vita contemplativa*" that frees movement from "having a goal" without toppling over into "destructive dispersion."[55] When we hear such anti-technological arguments for a return back to a rural, contemplative life, we are reminded of Benjamin's critique of the protofascist occultist Ludwig Klages as a precursor to eco-fascism. Despite his warnings, Benjamin did not simply reject Klages but rather attempted to create an alchemical constellation between Klages's notion of the *Urbilder* and Marxist dialectics.[56] Likewise for us, we cannot simply or easily turn off the Internet or unplug our cyborg selves from the postdigital networks that now define postmodern life. At the same time, we cannot simply accept the infinite flux and propulsive velocity of whizzing as a permanent condition, especially when whizzing dovetails so easily with right-wing accelerationism.

As a pataphysically impossible solution to the problem of movement, we turn to Agamben, who obliquely connects the s-pacing of drift in the studio with the temporality of *festina lente* or slow haste.[57] Although contemporary practitioners of research-creation such as Manning have highlighted the importance of moving slow in order to explore transversal resonances between art, politics, philosophy, and science,[58] we would add that the temporality of pataphysical alchemy in the studio has urgency to it and thus has a sense of quickening. Perhaps the most precise description of this strange temporality is found deep in our crypto-genealogy is offered by the Florentine painter Cennino Cennini who, in his influential *The Book of the Art* (c. 1400), outlined several different modes of temporality, including historic time, stages of training, mealtime, open-ended time, personalized time, calendar time, me time, and leisure time. The time of the studio is captured in Cennini's notion of "open-ended time," which seems divorced from ends while still having a sense of structure, significance, and urgency attached to it (and thus is not simply unstructured leisure time).[59] It is a time of extreme concentration and simultaneous casualness, of purposiveness without a purpose or destiny to be fulfilled. For us, this temporality is embodied in what we refer to as studious drift. Drift, therefore, is different from whizzing. Whereas whizzing simply has one dimension—the flash of a series of instances—and thus lacks duration, drift inserts a sway into propulsive acceleration, slowing it down so that lingering can happen, thus extending the instant, giving a modicum of duration to movement in the form of anarchival heaps. Drift concerns the friction of forces that curve or distort trajectories and thus inserts a gap (sometimes imperceptible) between a movement and its destination. Instead of reaching a destination, drift forms clumps out of detours. In this sense, drift retains something of whizzing's aimlessness. Drift is open and receptive to detours, yet this oblique or queer entanglement of lines does not simply unravel, gathering speed, hurling into the oblivion of nowhere fast. Instead, the interplay of rules and drift that generates and dissipates clumping masses introduces a certain amount of drag that slows propulsion, potentially causing drifting movements to loop back on themselves, turning them around rather than spiraling them out. The "destination" of drift thus inverts itself: turning course back towards its own potentiality for movement before that potentiality is oriented toward a final destination. The alchemical recipe and the patarule sabotage gestures, practices, concepts, objects, suspending their everyday functionalities. By doing so, these rules unleash a swarm of drift, but at the same time, the very same rules act as drag against such drift turning into whizzing.

Patarules and recipes set us adrift from the acceleration of linear marching and the rush of indiscriminate whizzing so that unexpected drifts can happen, unexpected things can unintentionally clump together. The studio is a space-time machine that makes another kind of movement possible, one that is paradoxical as it does not fit nicely into the linear story that moves from modernity to postmodernity. Instead, this time-space is oblique, queer, untimely, perhaps characterizing a *patamodern* dimension within life that falls outside of any chronological periodization. This is the time necessary for things to heap up, pile up, linger for a while before slowly yet urgently dispersing outward according to the ritualities of their internal rules.

Mapping the S-Pacing of the Alchemical Studio

Mapping drifting movements, rules, and drifted clumps in the alchemical studio is a topological problem that can be addressed with reference to knot theory. Indeed, the contact between the speed of movement, constraints, and spatial distribution forms ouroborotic loops. Somehow, the end finds itself beginning, looping back on itself, forming the alchemical ouroboros. According to Douglas Hofstadter, a strange loop is a cyclic structure that moves through a hierarchical system only to find itself somehow back at its origin.[60] Strange loops materialize not only self-reference but also paradox not unlike an M. C. Escher print in which the staircase upward only leads downward. The strangeness of strange loops is therefore more than a mere feedback loop. It moves through stages, levels, hierarchies, processes, only to discover/return to its own potentiality, as if such potentiality never left in the first place but was always equally present throughout all the ensuing stages. These strange loops, for Hofstadter, provide the essential infrastructure for all sorts of phenomena, including our sense of self. As such, strange loops are a basic aspect of the dark infrastructure of the universe, always withdrawing yet looped into phenomena as an immeasurable paradox. They guarantee that each loop will always drift away from logic and reason back toward a darkness that ensures the multiplication of pataphysical dimensions of experience. In this sense, ouroboros would not be a closed circuit so much as a repotentalization of a circuit that has drifted from its destination back to its own dark potentiality for magical/pataphysical alternatives always already present in the push-pull between loops and drift. Inoperative works emerge out of these loops.

For example, at the end of *Dr. Faustroll*, Jarry writes, "And behold, the wallpaper of Faustrolls' body was unrolled by the saliva and teeth of the water./Like a musical score, all art and all science were written in the curves of the limbs of the ultrasexagenarian ephebe, and their progression to an infinite degree was prophesied therein."[61] In the first instance, Dr. Faustroll's body takes the form of wallpaper (perhaps meant to cover the walls of a studio? The studio covered in the skin of a pataphysician . . . the room suddenly "wearing" human flesh, but inside-out as the flesh does not protect the inside from the outside but rather exposes the outside while protecting the inside). This is a body that unrolls, meaning it is curled up, in a loop, that can uncoil in order to be chewed upon. Emphasis is placed on the loop of the body, its ability to twist in on itself, and its plasticity. This body has also taken on characteristics of a musical score within which all art and science are rhythmically included without distinction. Art thus loops into science and science loops into art through the structure of poetic rhythms. The very same body copiously includes both old age and youth, scrambling any notion of "growth" or of maturity as the dialectical overcoming of immaturity. This is a body without distinctions, a schizophrenic body, or a body without organs that nevertheless finds a thumb to give it rules for unrolling. It is a queer body that drifts away from straight lines by drifting through curves and rhythmic oscillations that defy the laws of space—the inside of the mouth (teeth and saliva) becomes the outside of the room and the inside of the room becomes the outside covered in flesh—and time (as the progression of age ceases to make any sense). The loop of Jerry's body drifts. . . .

. . . and drift loops/twists. As the pataphysician Dr. Chubby Checker once sang "Let's twist again, like we did last summer/Yeah, let's twist again like we did last year/Do you remember when things were really hummin'?/Yeah, let's twist again, twistin' time is here." Twisting emerges through an intensified vibration (hummin) that is not measurable and yet produces the entwining of bodies (twisting). Twisting time is full of urgency, yet it must be done slowly enough so as one body can couple with another body forming a loop. Twistin' time is both the time of twisting and time as twisted according to a certain rule-as-paradigm that demands room for context specific singularities (dancing bodies under certain conditions, looping around one another). In this sense, Checker might as well have been describing the space-time machine of the studio (dance as studioing, or experimentation with bodies released from social gravity—not marching in unison or individually whizzing but rather twisting together). The twisting bodies on the studio dance floor thus embody a dark ecology of funky intensities between at least two bodies in entangled motion. For Timothy Morton, a dark ecology is a "strange loop"[62] of resonances that cannot be defined in terms of harmonious wholes or functional systems. We can add to Morton's description the strange loop between science and art, knowledge and pleasure that emerge through the (dance) studio. Twistin' time is here!

Loops can then tighten to form knots. Although Loveless titled her edited volume of essays, dialogues, and provocations on research-creation practices *Knowing and Knots*, the concept of "knot" is nowhere developed in any substantive way as part of a process of drift inherent to the movement of pataphysical alchemy.[63] For us, the centrality of knots in research-creation tie it (no pun intended) to a crypto-genealogy of alchemical and pataphysical studioing practices. For Deleuze and Guattari, a "line of drift" is composed of "different loops, knots, speeds, movements, gestures, and sonorities."[64] Drift loops in on itself but it also drifts into other lines and starts to produce clumps or heaps. Clumps temporarily slow down the pacing of a loop to form a spatial pattern defining the rituality of the space-time of the studio. Inoperative works are loose knots that, when clumped together, form the stuff of the studio. And in this sense, the *studio is a knot of knotted stuff*. The studio is the knot joining or weaving together various speeds, intensities, and affective resonances into an experimental heap. Studious drift pulls together a motley knot of knotted stuff (a clump of composed/composting stuff, or strange heap of inflection points), while the very same drift ensures that the clump is already drifting away from itself toward whatever it might become. Loops that stabilize enough to form knots offer up ecologies for dark precursors to arrive and pay a visit.

The origins of the idea of the knot reside in premodern, esoteric beliefs in the interconnectedness of all things (as above, so below). The cover of Athanasius Kircher's seventeenth-century book *The Magnes sive de Arte Magnetica* includes an emblematic frontispiece showing various kinds of knowledge (astronomy, philosophy, optics, music, theology, medicine, and so on) all linked together. The emblem is accompanied with the phrase "All things rest connected by hidden knots." The "hidden knot" was, of course, God, but it was also observable in the natural world in terms of magnets and their invisible powers to draw things together.[65] Whereas modern science concerns dissection and analysis of parts (an algorithmic separation and compartmentalization of the world into digitized pixels), premodern, esoteric knowledge involved discovering loops and contact points that could then form into knots.

A knot can also be thought of as a diagram. Commenting on Foucault, Deleuze argues that a diagram is a "spatio-temporal multiplicity"[66] that networks together or creates relay points between discursive and non-discursive formation. Whereas a structure is rigid, closed, vertically articulated, and inflexible, a diagram is composed of "alliances" that "weave a supple and transversal network"[67] within an open and "perpetual disequilibrium."[68] Diagrams are knots or weaves of ways of thinking, speaking, acting, and producing that never permanently solidify into a stable equilibrium. There is a drift that happens within the knot and across the knot as it can suddenly become entangled with other knots, forming larger diagrammatic webbing. They are, in other words dynamic, lacking a center around which they can orient themselves (there is no God as transcendental, vertical focal point here!). Deleuze concludes that "there is no diagram that does not also include, besides the points which it connects up, certain relatively free or unbound points, points of creativity, change and resistance. . . ."[69] In this sense, the diagrammatic knot does not prevent the clinamen as an "unbound point" of change from swerving into its field of elements, inducing drift. The clinamen as dark precursor exists as part of the outside of the diagrammatic knot as one of the "unformed elements of forces"[70] that has certain effects on how the diagram can assemble its elements.

Research-creation in alchemical and pataphysical studioing is a composition of loops to form knots out of otherwise dissimilar subjects, objects, and practices. Donna Haraway does not speak directly of alchemy or research-creation, but her theorization of SF gestures strongly in this direction—especially in relation to her analysis of "art-science-activist worldings"[71] such as the Crochet Coral Reef project. For Haraway, SF is a tentacular form of thinking that includes science fiction, speculative fabulation, string theory, speculative feminism, and science fact. She describes this entangled web as a list that "whirls and loops,"[72] revealing how science fact needs speculative fabulations and vice versa in order to create stories that generate response-abilities for multispecies living and dying together. Research-creation, on this account, is outside of settled categories, binaries, and functions and instead is a composition or open-system/diagram (a messy clump or drift, as we would say). Haraway gestures to the game of cat's cradle and Navajo string figures as embodied practices that generate precarious, networked patterns, necessitating new kinds of storytelling that are neither secular nor sacred, neither fact nor fiction, neither strictly knowledge nor pleasure. The subsequent stories she tells in her book *Staying with the Trouble* are tracings of how "knots knot knots"[73] or ways in which loops generate patterns in becoming that render capable mutually constituting partners (human and more-than-human, organic and inorganic actors). These are dynamic diagrams of "loopy tendrils" that "continuously spin"[74] and thus, are diagrams that are not spatially or temporally bound, containing the potential for surprising change and transformation. Knots are not closed, but rather tentacular insofar as the word *tentacle* comes from the Latin *tentaculum*, meaning "feeler" and *tentare*, meaning "to try."[75] Each knot feels its way, tentatively, precariously. Knotting knots out of loopy tendrils is a risky endeavor without guarantees or preexisting models for what constitutes "success" or "failure." SF is thus an experiment in research-creation predicated on knotting practices, languages, and actors into a string figure that is always already open to the arrival of a dark precursor that can introduce a new line of drift within an established pattern. String figures drift. Strings accumulate (knot) and disperse (loopy, diagonal, queer, strange trails that move without moving) at the same time. This is a clumpy, knotty, and drifty ontology of becoming.

For us, the alchemical studio is a diagrammatic knot that emerges through the loops of drifting discursive and non-discursive forms, practices, affects, and knowledges that simultaneously produce and enter its space-time. Indeed, Haraway's examination of the PigeonBlog performance/experiment that knotted together artists, activists, communication technologies, urban sporting birds, and city dwellers to map urban air pollution is not only research-creation in action but also a multispecies enactment of drift capable of transforming a city into a massive alchemical studio that experiments with what can and cannot be sensed/made/enacted collectively as a clumping swarm. The experiment threads the needle between science and art in order to set the city adrift from its business-as-usual division of labor between scientists and citizens, human animals and nonhuman animals, while also generating its own, unexpected clumpings/knottings that fall outside recognizable categories of either scientific knowledge or aesthetic pleasure. Such knottings can tighten (withdraw into a secret) or can suddenly loosen (opening up to new kinds of clumpy relationships with the outside, becoming a secret that gives itself away). The space-time of the studio is thus never fully closed, localizable, and instead is open and pliable. The alchemical and pataphysical studio is but one iteration of Jarry's time machine, which is itself an SF diagram: both speculative fiction (an impossible solution and exception necessitating rules-as-paradigms) and science fact (based on rigorous, algorithmic rules of physics).

Impractical Practice

Drift and recipes/patarules thus form knots that do not allow the movement of drift to become either monotonous marching (co-opted and given a direction, telos, and homeostatic predictability as in the laboratory) or flutterly whizzing (rapidly dissipating into protofascist accelerationism without ever clumping or looping back on itself). Both, in a certain sense, sacrifice drift on the altar of the empire of algorithms as algorithmic uniformity, efficiency, and thinness support both modernist marching and postmodern whizzing (which is never truly free but rather secretly guided by algorithmic calculation and prediction that only give the illusion of discretion). Whereas Han seems to want a return to the premodern rule-as-paradigm (a thick rule), we opt instead for a patamodern third option that is not reducible to one or the other but exists in the state of exception between the two: *an algorithmic yet non-calculative rule that is exact in its inexactness*. Such an alchemical pataphysics (or pataphysical alchemy) covertly smuggles in a minor difference within the algorithm so that discretion can re-emerge—a rule that is unruly in its invitation to swerve the knots that knots knot. The alchemical recipes and patarules of studioing thus protect the inefficient, slow haste of drift, making sure that the drift (of desire, inoperative use, and inflection) remains open to the close encounter with dark precursors.

To summarize, the alchemy studio is the spatial and temporal knot (of knots knotting), looping together patarules, alchemical recipes, and drift to form inoperative clumps (works). Regarding her own studioing practice, Manning argues that "ritualizing activity" activates a swerve "for the countable to the more-than"[76] thus maintaining the time and space for impossible solutions, exceptions, and particulars (so many exemplars of pataphysical play). Importantly, she draws a minor distinction between ritual (as a set of inherited lines and orientations) and rituality, which "despite its habitual nature, is nonetheless capable of shifting the field of experience."[77] For us, studioing is a space-time inscription of such rituality: a looping together by mapping routes between drifts and rules as inoperative works clump together. Although discussing witchcraft, we find the description of casting protective circles offered by Pignarre and Stengers as applicable to the alchemical and pataphysical studio we have developed through this gate. They write, "circles are woven, circles unravel."[78] Witch circles are the spatio-temporal pragmatic inscriptions of rituality that loop together certain bodies, practices, affects, and knowledge systems under the call for protection against capitalist (and fascist) capture. The dark infrastructure of the circle carves out a space-time for slow haste through experiment with strange loops. And while these loops are temporarily stable clumps (knots knotting), the clumps are always already shifting, drifting away from the drift of their accumulation. They weave and unravel, drift into and out of orbits, always gesturing back to a latent potentiality that is more than any given clump (a clump is itself and more than itself . . . it is whatever it is . . . a true accidental form, as Deleuze and Guattari would say).

Drawing the circle of the studio is an impractical practice because it does not abide by laws demanding productivity and efficiency but rather by recipes and patarules that are emergent and wayward. It is a ritual insofar as it concerns loops, recursive gestures that nevertheless are open to perpetual swerve effects that make them strange (paradoxical). And it is purposeless insofar as its destination, functionality, and telos are never determined ahead of its use. The studio insists on being a place where pataphysical and alchemical experiments in new use of one's organs can take place (a s-pace for becoming-metallurgical).

The resulting impractical practices of alchemical studioing are, as indicated throughout this gate, particularly hospitable to the arrival of a dark precursor from within the contractions and expansions of loops, all knotted together to shape the s-pacing of a studio. The dark precursor is a strange visitation of difference that transmutes drift into d-rift, or a rift in the fabric of space-time. Once released from operativity by entering into loops, the potentiality of concepts, gestures, practices, and materials become receptive to darkness. Let us put it this way: *Research* emphasizes recursive, impractical ritualities of drift while *creation* emphasizes the rift that the dark precursor can cause, unknotting the knots that knot. Alchemy is ontogenic research-creation with the potentialities within inoperative works (clumps of deactivated concepts, gestures, practices, and materials) in order to create hospitable conditions for the creative, transmutational force of the dark precursor. Through slow haste, drifts d-rift in experimental s-pacing(s).

Research-Creation in the Learning Classroom

In this final section, we want to reflect on our own studioing practices as university faculty and administrators and how we have transformed our learning classrooms into studio space-time machines through the looping of recipes, patarules, and drift outlined in this gate to form collectively inoperative, non-functional works. In this way, we have attempted to produce conditions conducive to research-creation (which is dependent on the slow haste capable of both being receptive to the swerve of a dark precursor and urgent enough to make sure that the experiment is not merely a game but actually involves real risks).

For us, learning is far too often the imposition of marching orders onto the movements of education. Learning as marching concerns conformity to algorithmic rules for what counts as progress or regress and evaluation of movement in terms of progress or regress (pass or fail).[79] This rigidification of educational movement prevents any encounter with the dark precursor (or at least makes it increasingly difficult). The march of education is, simply put, inhospitable to the dark precursor as a strange visitor (UFOther, as one of us has written elsewhere[80]). On the flip side, it is often the case that we find students continually plugged into their phones, whizzing around on the Internet in a state of perpetual distraction, unable to slow down (decelerate). Here students are moving too fast to perceive the anomaly of a dark precursor, which might be nothing more than an infrathin inflection in what appears. Neither of these states seem to be particularly educational to us. As such, over the last several years, we have decided to make an effort to transform learning classrooms for preservice art educators into studios for experimentation. We felt this was an urgent project because even the space of the studio in educational discourse has been taken over and made operative through the learning logic of "studio thinking," which attempts to operationalize what happens in the studio to make it more efficient and effective in developing creative skills needed for creative economies.

As members of a college of visual art and design, our practice lies somewhere in the ambiguous zone of contact between educational sciences and artistic happenings, between preparing students to become art teachers by teaching skills and expertise and experimenting in what education can be beyond learning through perceptual twists and through the production of inoperative works. As such, these classrooms-as-studios often buck up against student expectations. They expect professional development and instead get unprofessional drift that begins to loosen their affective and intellectual investments in certain (algorithmic) formulations of what education ought to be, how it ought to function, how its effects ought to be calculated/assessed, and what purposes it ought to serve. Stated differently, the classroom as infrastructure to support learning (as in a "laboratory" model of educational science) suddenly transforms into dark infrastructure for alchemical conjuring for drift, loops, and knots around inoperative works, wayward desires, and unexpected inflection points in taken-for-granted rules for educational advancement, growth, and development. Unleashing the drift means losing the way structured by the practical and efficient algorithmic rules of learning in order to curve away toward impractical practices of collective study.

In the fall of 2022, we co-hosted a "research-creation happening" at our home campus that brought together philosophers, educators, artists, and designers. Our provocation was to experiment with education beyond the space and time of learning by proposing simple protocols (sets of recipes and patarules) that would induce various states of drift. For this event, we also created our own protocol, which was as follows:

> Ask participants to come up with an educational theme (vote on the theme)
> Break participants into groups and each group will come up with its own set of rules for experimenting with this theme, creating small studio spaces (one to four)
> Groups will rotate every twenty minutes from one studio space to the next
> Before rotating, each group can modify (add, subtract, substitute) the rule set
> The experiment will conclude when each group returns to the rules it wrote and completes its own (modified) protocol

This protocol constituted a grouping of rules of thumb that served to initiate and accelerate drift among participants, who were eased into a state of uncertainty as they began to take up the individual rules and experiment with what the rules afforded (even if these affordances were unprofessional, awkward, or maddening). The composition of the studio space was itself diagrammatic in so far as the initial protocol intentionally perpetuated new groupings that themselves continued to branch out in new lines of flight that were omnidirectional or tentacular. Algorithmic parody was a concomitant feature of the protocol design, as the experimentation on the educational theme suggested a problem that could be solved by recursive means, but the algorithm never achieved a "real world" solution (only impossible variants). Instead, inoperative works circulated and clumped together forming loose thematic knots.

Here, it is instructive to examine a specific response to the protocol. In one such instance, participants determined a single rule of thumb, which read as follows: *create/develop/teach a gesture of enough*

This directive was written on a sheet of white butcher paper covering the table at the participants' studio space (tables at each station were covered similarly, allowing participants to write down their rules for public display, and to encourage other sorts of writings and drawings). The rule is a simple statement with complex implications. The gathering of verbs—"create/develop/teach"—is common to pedagogy, but "a gesture of enough" sounds almost cryptic and poetic. Gesture can mean the physical movement of the body but also a symbolic movement expressed through language or another communicative medium intended to affect the perception of others. Likewise, "enough" suggests that which has come to a quantifiable limit ("we have enough chairs for the party"), but it also can mean a limit of qualitative satisfaction ("your friendship alone is enough for me") or a limit of tolerance ("I've had enough of your shenanigans"). The rule introduces loops wedding embodiment, intellect, and imagination with notions of capacity. Utilizing the most straightforward meaning here, one could imagine a person raising an index finger or motioning with their hand at the neck to indicate someone should "cut it out." Enough! These gestures are surely taught as we learn how to read the bodies of others within our social framework, but creating or developing similar gestures requires a novel twisting of body and imagination to introduce a new element into the lexicon of social behaviors. It is like the dancer searching the body's possibilities to find the movement that best expresses the emotion or concept in question. And what of unembodied gestures? How are these constructed?

Responses to the rule by other participants offer some answers. One mode of response was the creation of weirdly inoperative enigmatic objects. For example, a set of participants left a drawing on a single sheet of printer paper, featuring layers of large scribbles in a variety of colors that resulted in a field of atmospheric noise that extended to paper's edges. In another instance, participants used materials at hand to construct a small sculpture: A roll of painter's tape was placed flat on the table and the hollow filled with small, colorful, fuzzy pom-poms that rose into a little hill. The way in which these objects address the "gesture of enough" is left open to interpretation, but they do point to art's role as a conveyor of gesture. These are inoperative works. What are they doing exactly? Their meanings are unclear, their presence obscure. They are not really concerned with the formal principles and elements of art, nor are they representational in the traditional sense. The idea of "enough" does have an emotional charge to it, as relating to experiences of satisfaction and tolerance, so one could interpret the works as attempting to express such emotion, or the idea of it itself. As works of art, they are not particularly interesting, but their use as inoperative works enabled participants to contemplate their own vestigial potential—enough was never enough, necessitating a return or twist back to the concept at the very moment of its presumed exhaustion in the work. They are by-products of the becoming-metallurgical of a subject caught up in the push and pull between drift and protocol. They are inflection points on the curve defining the multiple meanings and resonances of "enough." There is some darkness in these works as they slowly clumped together (never enough but also just enough).

Another mode of response did incorporate the body as a maker of gesture. A different set of participants left instructions on the table indicating that others should "sit on a chair and put your hands and head on the table for enough time." These directions were accompanied by a drawing—silhouettes of a pair of arms indicating where one should "put your hands" when seated. The gesture has a couple of connotations. One thinks of a punished child told to "go think about what you've done" putting their head down on their desk, or someone otherwise meditating or praying. Either way, the purpose of such contemplation is left open, as is what is meant by "enough time." What is a suitable duration, and who defines it—an authority of some kind or the seated individual? Perhaps part of the point is that "enough" is an arbitrary determination, or that it is situational, dependent on a constellation of contextual features. This response establishes a new rule of thumb that subverts the quantitative measurement of time by problematizing our ability to decide just what it means to be "enough." Even if we understand it as that which can be tolerated per individual, we do not know what informs the limit of tolerance. Patience is exhausted at different rates in different people, and for reasons that may be bodily ("I can't comfortably sit like this anymore"), emotional ("I feel like this is enough"), or philosophical ("What is enough, really?"). Again, this activity turns out to be an inoperative work revealing vestigial potentialities that were taken up and contemplated by other groups. The participants cease to function as agents who can confidently determine limits. They engage in an activity that would be considered useless by most standards, yet in doing so they avoid both the march and whiz of time, opting instead of a slow haste within the s-pacing of the experiment.

In addition to the responses above, each set of participants modified the original rule, generating inflections within the rules that unleashed new potentialities for drift. These included small adjustments to keywords, the addition of contextual language, and an expansion of the original sentiment:

> Create/develop/teach postures/acts of enough in various situations/activities/practices
> Develop/learn/teach a gesture/posture of digital learning
> List possibilities of enough in the room

While none of these modifications deviates drastically from the original rule, they do introduce some inflections worth noting. Gesture and posture both indicate the body, with an infrathin difference, as posture suggests an overall bearing or pose. It can also allude to an artificial attitude or stance, or an appeal to gain attention, as when someone is said to be "posturing." The performative nature of classroom instruction of course invites various sorts of posturing, which are used to get across an educational point. For example, one can think of an instructor playing devil's advocate during a critical discussion as a type of posture. Defining a posture of enough or that of digital learning is more difficult though, requiring a novel movement from the abstract idea to concrete interpretation through bodily positioning or artistic and/or intellectual expression. The modification almost presents itself as a dare—can one even come up with such postures? Again, precise answers are not as important as the subject achieving a state of aporia that provokes the recognition of inoperative potential. Achieving this state calls into question core assumptions about education, knowledge exchange and acquisition, and what counts as a valid educational output. The utility of this, if one can call it that, is that the subject gains insight into the dark infrastructure of the occasion of education and its effects. What does it take for something, *anything*, to become educational? What do we do with educational outputs that defy our usual taxonomies, metrics, or language of explanation? What does their mere presence indicate about the scaffolding for learning we construct? The creation of inoperative works marks a zone where progress and regress are beside the point, where the mercurial—in both common and alchemical senses—is given privileged space. What would normally be set aside is instead taken up for radical experimentation. The creative act, detached from a system of ends, breaks open, spilling over every available dimension.

The third modification takes the original rule and unleashes its full absurdity. "Possibilities of enough" is a confounding phrase, one suited to poetry, in so far as poetry is an accommodation to the fluidity and paradox of meaning. A list of such possibilities might include:

> The shadows that fill a cabinet once its doors are shut
> Dancing to your syllabus
> The sound of uncapped markers on the table slowly drying out
> The schism between a question and an answer
> Using your hand to count the number of fingers you have on your hand
> The window overfilled with reflections
> Gathering all the empty space in the room
> The taste of the last word that was spoken

In fact, a list of this kind would not look dissimilar to Duchamp's enigmatic series of (definitions) of "infrathin." But the eager learner will ask what's the point? We would argue that the modification recalibrates the subject's relationship to the topic at hand, forcing them to examine how it is that concepts get established to begin with. What is the enoughness of enough, its quintessence? There always remains an ineffable character to even our most defined concepts. The knot or clump of definitions is never tight enough to prevent further wayward drift in the form of modifications. The modification forced the participant to use language to solve language's central problem, which is its inability to speak to everything with absolute clarity. The list could be empty, or it could go on forever. Each response would be valid, and each would contribute to both our understanding of enough as a concept, and our inability to fully define that concept. At the edges of concepts lie affective intensities, wayward desires. We *feel* what enough is, even though we may not be able to state it. This feeling is not meaningless. It is an intensity that signals a knowing of a certain kind, but one that is not communicable through common means or the typical learning apparatus. To *feel* enough is to be put into a state of exception, the dominion of pataphysical affects and desires that are palpable yet inoperative. Enough is not *not* knowledge. Instead, it is the kind of (non)knowledge that is excreted from the contraction of science and art, pleasure and knowledge, and is often disavowed by learning, which is simply the measurement of knowledge accrued (often at the expense of the pleasures/sensations/affectations of wayward processes of experimentation).

The rules and subsequent inflections formed loops between pedagogy, art, language (verbal and body), and philosophy to encourage an intense engagement with the concept of enough. The drift initiated by the call for an educational theme in the original protocol turned back on itself in a recursive algorithmic parody, eroding the traditional sense of what counts as educational thematic. A swerve presented itself at the outset, as participants understood "gestures of enough" to be an acceptable educational theme. But how does it fit this description? One can imagine all sorts of examples of standard educational themes—assessment, study, progress, learning, equity, lessons, etc.—but "gesture of enough" is an outlier or exception. The participants have already hit an inflection moment at the outset, turning the rote notion of an educational theme back in on itself, exposing the inoperative dimension of its meaning. Inoperative works were formed as loops, discursive and non-discursive, drifted into one another, encouraging the slow haste of studious experimentation with the gesture of enough that was never quite enough to prevent further drift. Additionally, as participants moved from one protocol station to another, modifying rules of thumb as they made their way, new loops emerged amongst works. The manifested studio space-time became a diagram of entanglements, an ouroborotic knot that did not simply repeat its origins (swallowing its tail), but rather repotentialized these origins by setting its own rules of engagement adrift, inserting inflection points, generating swerves in the relationship between algorithmic exactitude (a formula for capturing enough) and paradigmatic discretion (aesthetic judgments concerning what counts as enough in the first place).

Let's twist again . . . or have we had enough?

Gate Three

Studious Risks and Research-*Creation*

As we have argued, studioing is a process of making loops: Drift and rules interact to form loops, and loops knot together into inoperative works, and inoperative works knot together to form the clumps defining the space-time coordinates of studios. If the resulting knots become too tight, they shut out the dark precursor (white cybernetics, mastery, surveillance, and governance over darkness are all *tight knottings*). The knots must remain loose. Loose knots are the topographic signature of ouroborotic entwinings, hospitable accommodations for wayward swerves.

This means that a studious knot is equal parts constraint (a certain amount of precise control over variables, a tactical clumping of elements, concepts, stuff) and risk (an openness to visitations from dark others). Interestingly, Anne Dufourmantelle describes risk as a discrete decision that decides the future, and in so doing reanimates the past, giving it a decisively new curvature. Drawing an analogy with sound, Dufourmantelle states that risk produces feedback. Sound "loops back" and "provokes a kind of secret intelligence that alone, perhaps, is liable to disarm repetition."[1] Risk is therefore not merely a break with the past, a leap into the future. Rather it has the structure of a loop in which the leap into the future reanimates the past, or rather a *difference* within the past. Unlike neurosis in which the future is subjected to the absolute law of the past, with risk, there is room for a minimal displacement that breaks open a new line of thinking, being, acting, making in the present. At work here is a "secret intelligence" that is esoteric, seeming to come from outside the subject—a dark, impersonal intelligence that draws out queer lines of becoming from the sedimentations of the past. The loose knot of studioing d-rifts, meaning that the drift of risk produces a rift with what has come before (even if this rift is infrathin), making time for creation.

For Michel Serres, one only becomes a subject by risking the parameters of subjectivity through mixing. He writes, "If you want to save yourself, take risks. . . . I am, I exist in this mixed contingency that changes again and again through the agency of the storm that is the other, through the possibility of his or her existence. We through each other off balance, we are at risk."[2] The self *is* the embodied sensation of being "off balance" through the mixing with otherness. To retreat from such mixing is not to fortify the self against the world but rather to destroy the self. The self is not anything other than the risks that put the self at risk in the first place. The resulting self is the historical accumulation of its risks. Like Dufourmantelle, for Serres, this self has a knotted structure. "We caress each other along our contour lines, we leave each other with various knots, in embraces that have changed shape."[3] The self is a tracing of a caress that has altered the shape of bodies as they unfold through one another and the state of consciousness that wrestles with and conceptualize such shape-shifting contingencies. Dufourmantelle's loop is a curvature of time whereas Serres's curvature is that of bodies. In both cases, temporal and spatial risks twist into loops and knots that undo what the self was and open it up to alternative becomings.

Turning to Erin Manning, we can more explicitly connect the theme of risk with research-creation. According to Manning, research-creation as studying with others in the undercommons is a "risk of losing our footing, risk of the world losing its footing, on a ground that moves and keeps moving."[4] In study (or what we would prefer to call studioing), risk predominates "the risk of the not-yet, of not knowing, of not even knowing how to know" and all we have is "study" or the "curious exploration of what the in-act can do."[5] Reading Dufourmantelle, Serres, and Manning together, studioing is the space-time incubator or alchemical catalyst for creative risk. The loose knots defining studioing offer a spatial topography for alchemical experimentation, inviting risk, which is connected to the temporal and spatial dimension of such experimentation (bending time so that destituent potentialities in the past swell up into the present to ignite a different future and twisting bodies toward one another to form clumps/activate the drift in existing clumps). Within the alchemical and pataphysical space-time machine of studioing, the *daemon* lives in shadows, curving the movement of scholars, artists, performers according to a secret intelligence beyond conscious knowledge or willful control. And this emphasis on risk, on *accentuating* risky experimentation, stands in contrast to the history of the laboratory, which concerns the calculation and minimization of risk—especially the darkness of risks. Indeed, Catherine Jackson's history of organic chemical laboratories in the nineteenth and early twentieth centuries indicates the central role of safety in defining this spatial-temporal form.[6]

Through this gate, we expand on the notion of risk and the loops it generates between past and future, self and other. In particular, through our crypto-genealogy of studioing practices we discover three very specific modes of risk: unprofessional risk, awkward risk, and finally, at its most extreme, the risk of madness.

At stake with each mode of risk is the arrival of a dark precursor that can generate just enough drift within an ouroborotic knot that creation is possible. Studioing is not simply research (re-searching, returning) but also *creation* through such rituality. In the following pages, we will explore three different types of risk—being unprofessional, being awkward, and being mad—before concluding with a consideration of the politics of such risk-taking. In sum, alchemical studioing and pataphysical experimentation concern risky practices in which subject and object are both rendered inoperative through looping ritualities, and this stands in contrast to the risks of contemporary creative capitalism and fascism, in which the former ends up not risking anything at all (thus perpetuating the status quo through management algorithms) and the latter risks too much, becoming a self-destructive, cancerous movement into the abyss.[7]

Risk of Unprofessionalism

Etymologically, the term *professionalization* comes from Middle English, from *profess*, an adjective meaning to profess one's vows. As one is transformed into a professional, one therefore professes skills and knowledge to others and vows to perform the profession in accordance with the highest standards (codes of conduct). In this sense, a profession is a secular form of a sacred vow. The notion of the sacred did not simply disappear in modern, capitalist society, but rather was absorbed into the sphere of the professions and professionalization. As Giorgio Agamben would argue, the secular is not the opposite of the sacred. The sacred acts to separate and divide the commons against itself, instituting a series of divisions into society between the included and the excluded. "Capitalism," writes Agamben, "generalizes in every domain the structure of separation that defines religion"[8] to the point where separation is fully absorbed into objects (the commodity is split within itself between use and exchange values) and the subject (divided against itself). Discourses and practices are also divided: head from hand, art from science, pleasure from knowledge.

Yet professionalization as the secularization of the vow has a major flaw. As Agamben argues, when one takes a vow, one simultaneously enacts blasphemy against one's self. Agamben breaks down the vow into three major component parts: "an affirmation, the vocation of the gods as witnesses, and a curse directed at perjury."[9] We find evidence of this claim in the very form that many vows take, "I swear to God, if I don't do x, I will go straight to hell!" or its more secular version, "I will do my duty or resign, so help me God!" Vows and oaths are therefore predicated on cursing the self. Vows and oaths do not abolish blasphemy so much as incorporate blasphemy into their form. At the heart of each professional vow there is an infernal contradiction that prevents the vow from ever securing the sacredness of the profession, hence the need to continually take more and more vows, profess one's skills over and over again. The vow taken is never enough, and one can never profess too much to one's profession precisely because at the heart of such a vow is a blasphemy against the self, a negation of the sacredness of the vow that inserts an infrathin difference between the fallible individual and the sacredness of the profession that they are meant to uphold and fully embody. Stated differently, the sacredness of the code of conduct does not protect the profession from blasphemy but rather lives on and through blasphemy, producing the need for incessant vow taking and professing without end precisely because the vow *cannot be fully believed, cannot be fully trusted*. And if one does not abide by the command to take a vow and profess, then one is abandoned by the profession as unprofessional. Ironically then, the only way to *not* blaspheme one's self is to prefer not to take the oath of the profession in the first place (a pataphysical inversion of Pascal's wager)!

For Agamben, profanation is a suspension of the internal dialectic of the vow of the profession. Instead of professionalization, Agamben would embrace the profane life of the amateur who loves knowledge without having it (and thus embraces a queer non-knowledge). To profane, according to Agamben, "means to open the possibility for a special form of negligence, which ignores separation or, rather, puts it to a particular use."[10] Profanation "neutralizes what it profanes" and "deactivates the apparatuses of power and returns to common use the spaces that power has seized."[11] Or we would say, profanation is about *creation*. Whereas the professions concern the baptizing of creative individuals, studioing concerns the profanation of individuals, ideas, acts, gestures, and so forth so that acts of creation can take back the places that the professions had seized for the exercise of their powers. The profanation of studioing is an occult power (a studious secret that *gives itself away*) within yet against the professions, subversively infiltrating from the inside in order to turn them inside out or upside down in the name of a new, unanticipated use that is no longer the exclusive right of those belonging to the profession (who have taken a vow).

The Finish artist Pilvi Takala's performance piece titled *The Trainee* (2008) is precisely an experiment with such unprofessional profanation. Takala demonstrates that studioing is not bound by a specific space-time, but rather generates a distinct space-time whenever and wherever laws of professional decorum are suspended through ouroborotic loops between drift and rules. For *The Trainee*, Takala attended a month-long "internship" at the global accounting company Deloitte under a pseudonym. After two months as a productive and professional trainee, Takala suddenly stopped working yet preferred not to leave. She would sit at her desk in silence. When others asked what she was doing, she would simply respond that she was engaged in "brainwork." The ensuing confusion was recorded via hidden cameras and emails that she collected and presented for *The Trainee*. The power of the vow to work and be productive (to be professional) was therefore profaned precisely by a trainee who is not a professional and not *not* a professional at the same time.

For us, Takala's gesture of "preferring not to work" while remaining at work transformed the corporate office space into a studio. It is a profane work that does not work. Importantly, Takala does not destroy the office or blow it up or even engage in overt resistance/rebellion. She merely inserts an infrathin inflection between professional forms of life and structures and their purported ends. When this happens, the forms and structures are opened up for new, profane uses beyond vow taking and professing. Takala has drifted away from professing while going nowhere, remaining firmly placed in her cubical, before her laptop. The performative studioing practice is a strange kind of time machine that inserts a different temporal spacing into the office, one that is no longer oriented toward a specific, predefined end measured by performance reviews. Profanation, stated differently, deactivates the underlying telos or means-end logic informing the forms and structures of professions, and through deactivation, one can finally play with the forms and structures as if they were toys or raw material for reinvention. This playfulness has its own temporal flow (slow haste) in which the "employee" falls off sync with professional marching orders. Through a simple rule ("preferring not to work at work"), Takala introduces a swerve into office life, a kind of dark precursor for which there is no answer or explanation. We might even go so far as to suggest that Takala's space-time machine generates an "upside down" office or bizzarro office around itself—a special kind of pataphysical state of exception at the very heart of professional practice. The "results" of such parodic practices are oftentimes frustrating for those who remain faithful to the sacredness of the profession, and thus believe our actions should abide by its laws and duties. Such individuals remain inside the gravitational force field of professional physics and refuse to drift into the alternative temporal and spatial coordinates of pataphysical studioing.

As a trainee, Takala occupies a limbo space between inside and outside, professional and unprofessional, studying and working. Because of this, she is difficult to "sacrifice" or merely kick out of the office. And in this manner, she is able to suspend the logic of sacrifice that exists within the secular professions as a religious inheritance. No longer a mere amateur and not yet a professional (who has taken the vow of obedience), Takala's performative studioing incubates a strange state of pataphysical exceptionality that seems to belong nowhere and is incapable of solving any problem (indeed, its mere presence generates unending problems), and for this reason is capable of embodying a destituent potentiality that is not destined for any particular end. Her brainwork is observably uncreative; she prefers not to engage in creative economies of the professions, yet this very same inoperative work manifests a destituent potentiality for creation as such—the appearance of something startlingly weird that ought not to happen when and where it happens, a true professional exception (or an exceptional profession, meaning a profanity).

In sum, Takala's performance is an example of how studioing need not be destructive or utopian (utopian in the "bad" sense of a blueprint of the future). Indeed, the space-time machine of the studio is *atopian*. She unlocks various potentialities for living a destituent, experimental life by suspending the underlying professionalization of the profession and thus rendering inoperative (even for a brief moment) the divisions that define the life of the professional within the creative economy. Such performance is research in that it delves into the hidden complex of rituals and infrastructures supporting neoliberal professionalism (not unlike sociological inquiry) while also creating a hollow within this space-time for something vestigial to make a sudden appearance (a queer pleasure or flavor that is not reducible to any knowledge that is produced as an outcome of the performance).

Risk of Awkwardness

In his brilliantly insightful and entertaining book *Awkwardness*,[12] Adam Kotsko develops a philosophy of awkwardness that is fitting for our increasingly awkward social reality. Much like the word *mistake*, awkwardness means wrong-ward or indicates something turning in the wrong direction. Keeping this basic definition in mind, Kotsko then outlines three kinds of culturally pervasive forms of awkwardness: everyday, cultural, and radical. These states of awkwardness are, for Kotsko becoming increasingly prevalent because of overall social, political, and economic instability brought about by de-industrialization, post-Fordist production, and the (more-or-less simultaneous) rise of civil rights movements. Symptomatic here are the radical upheavals of the 1960s and a concomitant erosion of "traditional values," which had provided a (repressive and exclusionary) *logo*. Without such a *logos*, social norms have become increasingly fragile and ungrounded, opening up an expanded terrain of awkwardness that is perhaps best captured in the dialectic between Trump's totally unhinged assaultive speech (which desperately attempts to thwart the awkwardness white males feel by a loss of *their logos* by provoking escalating fits of awkwardness) and the liberal attempts to regain some kind of civil discourse through a notion of "political correctness" (which also does not solve the problem so much as create awkward forms of backlash). We would add to this analysis how awkwardness is the phenomenological state that exists between risk-taking and creative outputs in the creative economy. The creative economy demands risks, but risks induce potentially awkward situations. Creativity is the evaluative matrix that enables awkwardness to be potentialized as desirable, profitable. But, as Kotsko argues, there is something within awkwardness that can potentially unravel this equation.

To begin, everyday awkwardness is perhaps the most intuitively simple of Kotsko's three variants, and suggests that the source of awkwardness in a given situation is the responsibility of particularly awkward individuals. Here, it is the individual who is awkward. To remove the individual then, is to solve the problem of awkwardness, and thereby return the situation to normal. To move from a wrong-ward direction to a right-ward direction, one can simply eliminate the bad apples.

Kotsko provides ample examples of everyday awkwardness in media culture, including the US version of *The Office*, in which individual characters are portrayed as causing workplace awkwardness (another hack of the primary site of creative capitalism akin to Takala's performance). For instance, the manager of Dunder-Mifflin paper company, Michael Scott, generates awkwardness through his hapless buffoonery, which is viewed with ironic amusement by more normal (non-awkward) characters such as Pam and Jim. There is an overall sense, that the awkwardness Michael produces could be solved if he were simply removed. Of course, there is always the risk that someone of equal or greater awkwardness might replace him, which is the case with the series, but the possibility remains real. For us, what is important here is how individual awkwardness demands creative solutions to office management. Thus, creativity becomes wedded to managerial surveillance and governance in order to tamp down the potentially disruptive aspects of awkwardness. Indeed, we would argue that the managers in Takala's office most likely viewed her through the ideology of personal awkwardness, and their various responses were all attempts to employ creativity to find a professional (meaning polite, law-abiding, institutionally legitimate) way to cope with her weirdness.

But there is a serious flaw with this concept of awkwardness. For Kotsko, awkwardness is first and foremost *social* rather than individual. Indeed, he utilizes Heidegger to argue that awkwardness is a special kind of mood that is shared or held in common with others. What makes awkwardness unique is precisely this intersubjective dimension. Unlike Heidegger's more famous analyses of anxiety and boredom, both of which are highly personal and suggest a withdraw from the social world, awkwardness indicates that one is always already in a social situation in which one cannot fully escape. One is always dealing with others, wrapped up with others, and as such, always exposed to the possibility of making a wrong turn or taking a mis-take that violates certain social laws, norms, and expectations. To think that awkwardness is simply the result of individual choices or dispositions is to miss that awkwardness underlies every situation in which we find ourselves as members of a shared world. Second, it also lets off the hook the social order itself. When awkwardness is projected onto an individual coworker, focus shifts away from the social laws as *producing* or *inducing* awkwardness. It is here that Kotsko turns to his second form of awkwardness: cultural.

Cultural awkwardness proposes that the problem is not individuals but the social system as such: its norms, laws, and values produce the problem of awkwardness, that, in turn, is leveraged by the system to continually promote the illusion of itself as perfectly functional and reasonable. Kotsko points toward the films of Judd Apatow for examples of such cultural awkwardness. In many of Apatow's films, the cultural and social systems are fundamentally broken, producing an awkward surplus, especially for male characters who no longer can believe in the ideology of the white, male, productive citizen, father, protector. Such characters are left without a firm cultural orientation to guide them in what it means to be a man, and thus continually fall offtrack, creating awkward situations for themselves and others around them. Yet the implicit critique of the culture as broken, flawed, awkward is often resolved in Apatow's films with a reaffirmation of the very cultural norms and values that they satirize. In short, Apatow's hapless male characters who actively resist becoming fathers, husbands, or productive employees, end up getting married and living happily ever after. Summarizing this recouperation of awkwardness by an already awkward system, Kotsko writes, "The way to lie with the awkward, broken system that married adulthood has become is to redouble the awkwardness by making the utopian awkwardness represented by male bonding the servant of the very order it thinks it's fighting against."[13] Awkwardness is rehabilitated back into a creative economy. In this sense, the risk of awkwardness is always already a calculated risk. In the promotion of creativity, creative economies allow for a certain amount of awkwardness, which is disruptive but only in the right ways at the right times so that it can be reevaluated in the long run as an investment that paid off. Risk, on this reading, is not simply an unfortunate, collateral side effect of creative production, but is secretly working to reestablish the very norms and values that it would appear to challenge. There is no risk that risk will not generate a happy ending, and in this sense, creativity can no longer create or generate anything new and whatever risk is left is fully domesticated. Hence the aesthetic enjoyment of Apatow's films: One can laugh at the awkwardness of the situations knowing that at some point all will return to normal. Creativity in such cases is the ingenious reconciliation of awkward waywardness with the (white, male) social order.

On this view, awkwardness is a means to an end: the stabilization of an unstable and increasingly ungrounded creative economy of entrepreneurs anxious to find solutions to its teetering economic precariousness of neoliberalism. But what would happen if awkwardness were a pure means rather than a means to an economic end? For Kotsko, this question pushes him to his next category: radical awkwardness. In a state of radial awkwardness there is no longer any set of governing norms or laws mediating behavior in a situation. As Kotsko summarizes, in such a state, "awkwardness [is] enjoyed for its own sake . . . becoming its own kind of grace."[14] Instead of norms against which behavior can be evaluated, judged, and ultimately included or excluded as part of a community, an awkward community appears that suspends and renders inoperative such standards of measure. Radical awkwardness exposes individuals to one another directly and without mediation. In what might strike some as an unusual move, Kotsko turns to the television show *Curb Your Enthusiasm* and the caustic character created by Larry David as an example of radical awkwardness. For Kotsko, the underlying philosophical importance of this show is the new theory of political community that it proposes: "That no one should be forced to conform to the arbitrary social norms of others. . . ."[15] Larry David, at his best, does not simply (a) reduce awkwardness to individuals, or (b) attempt to prop up social norms by rehabilitating awkwardness as part of its ideological support. Instead, he merely *dwells* in and *enjoys* awkwardness as an open-ended question that is shared, that is held in common. Indeed, this silly and irreverent show actually reveals how awkwardness as a mood has a certain ontological priority over anxiety and boredom insofar as it reveals how we are always already together in a state of ambiguity that laws and social order attempts to negate, deny, or repress (as much as possible).

At this point, we can add our own example, this time taken from the field of dance: the social choreography of Michael Kliën and his unique rethinking and reimagining of the "dance studio." For his piece titled "Parliament" (ongoing), Kliën gives his performers minimal rules of engagement. He takes a group of strangers (mixing dancers and non-dancers) and places them in a large, closed room without furniture for six hours. During this time, they cannot speak to one another. These two rules open up a highly experimental zone in which participants begin to play with gestures and modes of embodied communication. Some participants sleep on the floor, while others engage in any number of idiosyncratic movements, sometimes in isolation, but more frequently together in small groups. Others begin to groom one another, combing hair with hands or giving massages. The point is that the rules enable a certain amount of body drift to happen that renders inoperative the ways in which bodies are supposed to relate to one another in common, everyday situations. The speed of bodies is interrupted. Whereas bodies often whizz (online or from one job to the next), here bodies slow down. The pacing of the bodies seems to be simultaneously urgent (meaning important and significant) yet also slow and pensive. The bodies are figuring out how to negotiate with one another within the suspended state. These are not bodies that march together in uniformity. Instead, bodies roll about, twitch, sleep, sprawl, crawl about—they form clumps. And if they do engage in marching, it is a strangely parodic affair as if the marching itself is no longer symbolic of communal conformity. In other words, there is an infra-thin division within the most familiar looking gestures that makes them weird and uncanny. And the spacing between bodies is displaced. The typical distance between strangers collapses or exaggerates as bodies take up postures that are unusual and would otherwise be considered uncomfortable in everyday life. Bodies are no longer propelled by clearly defined, average, everyday objectives or ends. These ends have been neutralized by the basic rules of engagement, generating gestures that might look familiar yet are not oriented in the right ways, at the right times, to the right things. Bodies adrift lack ends, yet there is still urgency to the movements. But this urgency is not directed toward fulfilling a specific potentiality. Instead, the urgency twists back on itself, becoming an ouroborotic loop in such a way that the body can rediscover its destituent (undestined) potentiality for awkward, collective movements.

In short, these are bodies that are experimenting with s-pacing, or are generating awkward s-pacings for which there are no predefined, preexisting rules of engagement. Instead, the rules have to emerge out of the drift and are fragile and fugitive, uncertain and ungainly. These bodies experiment with unleashing gestures from typical places and functions in order to try out new uses, new possible ways of defining what bodies can do collectively in a common space. They risk awkwardness, or wayward, queer twists and curves that might not be understood or recognized. Kliën's title "Parliament" conjures up a politics to come, or a politics without a clear destination/end, without predetermined laws looming above, and whose agent is a coming community of strangers that share nothing more than minimal rules for inventing an alchemically transformed collective body. The results are as comical as they are mysterious, reminding us of the anarchic depictions of alchemist studios with which we started this book.

The awkwardness of "Parliament" is cryptic and occult yet also public. Unlike the studious secret of alchemy, which moves from the inside to the outside (to become part of the commonwealth), here the movement is from the outside (the collective, the public) to the inside (of a rule-bound studio space closed off from the wider world). This is like a "controlled" experiment in which variables have been removed and social performances stripped down to certain, minimal possibilities that then open themselves up for elaboration. This reduction or miniaturization of the social enables a unique brand of research-creation that equally investigates the meaning of the political while also actively creating preconditions for the arrival of a not-yet-become anarchic political horizon. Stated differently, "Parliament" is a collective effort to research what is possible within a pataphysical state of exception in which most, if not all, rules of social tact have been suspended and all laws neutralized, allowing bodies to risk the awkwardness of drifting into one another (or away from one another, as the case might be). The collective, insular nature of this particular risky venture can be compared to the individual, professional risk of Takala, who remained isolated and exposed. And as we all know, unprofessional professionals risk sacrifice and death, as was the case with Bartleby the Scrivener.

Risk of Madness

Deleuze warns that schizoanalysis should not be equated with clinical madness, and yet the drift induced by studioing can sometimes tilt the scales. The difference between clinical and critical schizophrenia is infrathin. On the one hand, neurodiverse freedom is an awkward freedom because we do not know now to think or feel and the "typical" ratio between knowing and enjoying can shift parameters dramatically.[16] What we call neurodrift is a loosening of the tightening cognitive-perceptual knots that bind the body to certain ways of sensing itself and the world, or tie thought to common and good sense. Against such normative claims over the self, Deleuze playfully asks, "should we go a short way further to see for ourselves, be a little alcoholic, a little crazy, a little suicidal, a little of a guerrilla—just enough to extend the crack, but not enough to deepen it irremediably?"[17] The qualifying statements "a short way further" and "just enough" indicate a certain amount of discretion and the need to set certain rules of thumb for pulling back on drift (so that it does not fully lead to a crack up). To prevent the most dire of consequences (suicide, madness, alcoholism), Rosi Braidotti argues for an ethic of sustainability as a "thin barrier against the possibility of extinction."[18] Again, the emphasis on infrathin barriers that are so thin as to be transparent and almost imperceptible is an important dimension of any alchemical or pataphysical experiment. Otherwise, neurodrift can spiral outward to the point of total deterritorialized explosion, a radical cracking up in which the subject is destroyed by an intensity of affect that is unleashed by a dark precursor surging over and through a body that becomes increasingly twisted in on itself. Thus, we end with a case that demonstrates the *profound* risks of tarrying with the dark powers of studioing.

When Leonora Carrington entered the sanitarium for a mental breakdown after the imprisonment of her lover Max Ernst, she was confronted with a terrifying imprisonment. She could not tell if she was in a "hospital or a concentration camp."[19] Her doctors and nurses subjected her to unthinkable tests, strapping her down to the bed, force-feeding her, and giving her a series of injections, one of which produced a contusion on her leg that nearly crippled her for weeks. Yet despite the ordeal she suffered, Carrington was able to transform her impoverished room into an alchemical studio. After a series of injections, she was finally given back the simple objects that had been confiscated from her upon her entry. With these objects she started "combing solar systems to regulate the conduct of the World."[20] Several French coins became transformed into symbols of the downfall of humanity (a critique of capitalist greed). Her red-and-black refill pencil became an obelisk of Intelligence. A jar of face cream with a black lid became night, the other with a green lid became the Sun. Her nail buff, shaped like a boat, became a vessel for transporting her into the Unknown as well as a "talisman protecting that journey."[21] After several other items, the list concluded with her Tangee lipstick, which came to represent art. Carrington recalls, "Happy with my discovery, I would group these objects around each other; they wandered together on the celestial path, helping each other along and forming a complete rhythm."[22] And in this way she "gave an alchemical life to the objects according to their position and their contents."[23]

These rituals enabled Carrington to transform the room of her imprisonment into a secret chamber full of meaning in the heart of an institution that resisted meaning. What happened inside those walls was precisely the terror that darkness can bring: darkness of the camps and of torture, a darkness that is imposed from the outside on those who are most unfortunate. Yet in this darkness, Carrington was able to sensitize her eyes and her sensorium in such a way as to detect another darkness, not so much a light as a dim, shadowy force (a dark precursor) emitted not from the sun but from the strange place she described as the "down below" or a subterrain paradise.

At a certain point, Carrington received a second Cardiazol shot. She closed her eyes to shut out the pain, but with eyes closed, she had a vision of a door to the down below. Wandering down an alley and between trees, she went into the sanatorium's garden where "everything became richer and more beautiful."[24] At this point, her mind drifted into an intense vision of a multistoried building, which she entered to perform various rituals. "I opened all the windows" she recalled, "as I would have opened those of Consciousness, except for one: the mauve one, the Moon's, since my 'moon cycle,' my menstrual period had stopped. Having concluded the Work, I walked down the stairs and returned to 'Egypt.'"[25] Through complex, exacting ritualized rules that dictated the arrangement of objects and her precise movements, Carrington traveled without going anywhere. She was able to unlock surreal dimension of a sterile and seemingly hopeless place and thus to enact a visionary escape. The grounds of the sanatorium became enlivened with esoteric meanings that enabled her to alchemically transform herself from a patient into a pilgrim. Of course, these alchemical rituals were "impossible solutions," but this is all that is possible in an impossible situation in which one has no agency or control over one's life. Instead of extinguishing the darkness, Carrington pushed into it, drifting into the down below.

Yet, for Carrington, her experiments with alchemical studioing dipped too far into madness. Thus at critical moments, Carrington realized she needed to free herself from these alchemical visions and from the multiple personalities that had come to occupy her mind. Neurodrift had become the onset of clinical schizophrenia. She lamented, "I had to get rid of everything my illness had brought me, to cast out these personalities, and thus begin my liberation."[26] The drift overwhelmed the *discretion* associated with rules of thumb. The risks became too great. She lost sight of the collective experimentation that is awkwardness, becoming increasingly isolated. Beyond the appearance of unprofessionalism, her rages against the institutional apparatus ultimately turned against herself as well. And finally, the risks exploded into full insanity, which compromised her health. Neurodrift accelerated to such a speed and intensity that Carrington was no longer able to sustain it.

Disappearing into the darkness of the down below, Carrington also lost contact with her works. Only later, after being released, did Carrington write her memoir and paint her portraits, enabling her to contemplate the potentiality of drift rather than being completely overwhelmed by it. Through her artwork, the alchemical work upon herself was given a form. The two created a knot that could loop into a ritual practice, granting relative stability and rhythm to her drift. Thus, Carrington's desire was not merely to abandon drift and return to normalcy (neurotypicality) but rather to find rules that enabled her to produce inoperative works that could be contemplated (studied). As Deleuze and Guattari warn, "Every undertaking of destratification (for example, going beyond the organism, plunging into a becoming) must therefore observe concrete rules of extreme caution: a too-sudden destratification may be suicidal, or turn cancerous. In other words, it will sometimes end in chaos, the void and destruction, and sometimes lock us back into the strata, which become more rigid still, losing their degrees of diversity, differentiation, and mobility."[27] Rules are there to slow the speed of drift down (to avoid whizzing). The rules thus enable us to have a certain cautionary discretion. At the same time, drift prevents the rules from becoming rigid (and the movement of studioing becoming a march). The two form a ritualized knot. Of course, the dark infrastructure of the sanitorium would not allow this, and she did the best she could. Indeed, her extreme madness must be seen as a result of certain medical infrastructures that would drive anyone insane, let alone someone as sensitive as Carrington. The point is that she did not reject the need to take risks, indeed her art testifies to her willingness to repeatedly risk madness. But she did so in a more convivially open studio that was hospitable to the kinds of rituality necessary for inviting dark precursors rather than in the fugitive studio, which she had to construct, alone, under extreme duress and under constant surveillance.

Dufourmantelle argues that the risk of opening the self up to a "liberating amplitude"[28] beyond the boundaries of the conscious self can result in madness or alcoholism or both. Visions, hauntings, and prophetic dreams are all manifestations of the effects of the dark precursor on the organization of perception and cognition in a subject that put "us" in "immanent danger."[29] For Dufourmantelle, as for Deleuze, artistic research-creation is one way to avoid total madness when one contacts the beyond. She writes, "Artistic creation is always at grips with the apparatus out ahead of you that informs you, somehow unbeknownst to you, and that unfolds on the canvas, in the score, or on the page even before consciousness takes notice. Only upon rereading does it become aware."[30] Importantly, the production of inoperative works can stave off madness (or at least its most dangerous aspects), allowing the subject to contemplate their destituent potentiality (or dark potentiality).

The Politics of Risk

In this gate, we have outlined three kinds of generative risk through alchemical studioing practices, each of which is hospitable to the curvature of the dark precursor, creating something new within situations that would otherwise be mere repetitions of the same (in the form of routinized labor practices, individual and cultural awkwardness, or institutional normalization).

But this raises a final, *political*, question for us: Should one risk everything?

To answer such a question, we will compare and contrast how risk appears within neoliberal capitalism and fascism, respectively, in order to argue the following:

> Capitalism: no risk at all (creative economies lack acts of real creation)
> Fascism: risks everything (and ends up in a state of destructive nihilism rather than creation)
> Alchemical research-creation: risks something (in the name of hospitality toward the dark precursor)

First, in relation to neoliberal capitalism, we find a confluence of risk management and creative economies. In the book *The Cult of Creativity: A Surprisingly Recent History*, Samuel W. Franklin[31] argues that creativity gained heightened social value in the West during the Cold War. It expressed American exceptionalism against the technocratic conformism of the Soviet Union. Although lacking firm psychological foundations, the discourse of creativity became increasingly important for businesses interested in innovation, not for national defense so much as for profit. More innovation demanded increased investment into research methods to pinpoint creativity in potential employees. The democratization of creativity—and thus critiques of the more "elitist" and Romantic notion of the genius—was decisively the product of capitalist interest in maximizing the development of more products, with the personal value of feeling that individual creatives are "self-actualizing" through creative expression. Everyone can become an artist or designer, and thus everyone's labor power can be commodified. In this way, creativity is a discourse that primes people not only to accept but also actively participate in and identify with the extraction of their ideas. Indeed, it is a sign of one's creativity to have one's creative output coopted by capitalism! The discourse of creativity gives added meaning and significance to otherwise banal, capitalist alienation.

What is most interesting in Franklin's analysis is the idea of the *cult* of creativity. Here we are reminded of Walter Benjamin's notion of cult value and its connections to aura.[32] Originally, aura related to individual objects or artifacts and their unique existences embedded within traditional ritual practices, yet within the creative economy, the aura seems to shift from fetishization of objects to fetishization of immaterial, soft skills such as creativity itself. The cult of creativity invests heavily in the aura of the individual and the educational development of the "creative individual" who is a value both to him/herself and to the broader creative economy. Aura does not disappear but rather is displaced from object to subject—neoliberal, creative economies reveal the sacred within the secular (as Agamben would argue). The capitalist entrepreneur as creative icon takes on a certain aura, or magical glow capable of conjuring the future and enchanting the masses. We only have to think of such figures as Steve Jobs or Elon Musk to understand the fascination the public has with these embodiments of creative aura, these (quasi)sorcerers of creative entrepreneurialism.

Creativity becomes an ideology of the creative economy, and the studio becomes co-opted as a place in which creative capacities ("studio thinking" skills) are developed as a marketable asset. In this sense, "risk" in the creativity studio relates directly to how well one's education does or does not help actualize creative potentials. Education becomes the fulcrum for the ultimate extraction not of labor-power but of creativity-power (in which one's aura is co-opted as one's unique "brand"). And if one does not maximize this creative potential by investing in the development of creative skills, then one is a failure and marginalized in an economy now driven by the immaterial labor of designers, programmers, influencers, and so forth. Thus, risk and creative economies come together. Creative economies pluralize risks while also attempting to manage such risks through ever more precise, predictive algorithmic rule sets. As Dufourmantelle argues, the current world is obsessed with both risky behaviors and, equally, calculation and management of such risks. Creative investments demand taking risks, and risks demand increasingly sophisticated, creative responses. Risk becomes an "unquestioned value"[33] within creative economies. But such risk ultimately *risks nothing* and thus *creates nothing* (outside of profit). Risk is always already managed (or manageable), and whatever creativity emerges is always already prefigured by the needs of the market. How might we once again take a risk with risk beyond calculation, prediction, and value production?

The stakes in defining research-creation as *creation* become clear: Research-creation is not a practice that develops and expresses creativity as an individual cognitive and emotional set of marketable skills. Instead, it is an *event* of d-rift that happens within the loops and knots of the dark ecology of studioing. It is the event of arrival of a dark precursor that punctures the subject's ability to orient and make sense of themselves and their actions. This is not the effect of a creative will to harness the powers of the cosmos for one's self-improvement or corporate profits but rather a willingness to expose the self to the risk(s) of a darkness that is beyond such control. While engaged in the pataphysical alchemy of studioing, the subject risks itself in and through the production of inoperative works that are traces of an encounter with a dark precursor whose secret intelligence does not abide by the logic of neoliberal, entrepreneurial creativity in which all risks must be made operative, functional, and oriented toward specific ends. Indeed, the subject of studioing becomes increasingly defined by the production of (an)aesthetic, vestigial organs, sensations, and affects rather than by marketable, "twenty-first-century" skills—limbs that do not obey, behaviors that are not disciplined/organized, gestures that are inefficient and extraneous to the "task" at hand. This existential risk is ignited through the production of inoperative works, or works that do not measure up to expected outcomes, and do not actualize a potential (for creativity) so much as throw into relief the indeterminate qualities of destituent potentiality as such (now released from functional ends as determined in advance by the creative economies that parasitically thrive on creative outputs). In this sense, studioing is an occult intervention into the cult of creativity—a fugitive, criminal act of creation against the creativity of contemporary capitalist, white cybernetic surveillance, control, and governance. If there is aura in studioing, it is an aura of blackness, a negative aura that is present in its absence rather than a white aura emitted by a self-proclaimed creative Prometheus.

In our three previous examples, we have highlighted how studioing induces unmanageable risks that fall outside the calculus of capitalist rules-as-algorithms, which simultaneously attempt to promote creativity while also calculating potential dangers (all in accordance with profit as the ultimate arbiter of meaning and value). To be unprofessional means that one's behavior is not "value added" to the teamwork necessary for sustaining creative economies. Likewise, to be awkward means that one's actions are no longer oriented in the right direction, in the right time, and in the right way. These are actions no longer oriented by and toward profit or social utility of any kind. Instead, actions drift away from the capitalist horizon that defines what is worth doing and knowing and instead drift into queer clumps of what remains in capitalism's wake (the accursed share of vestigial uselessness). And finally, madness disrupts the control of desire by introjecting an excess that cannot easily be recuperated by capitalist performativity (as mad and as maddening as it already is). As Carrington's life reveals, madness creates radical new meanings out of closed, disciplinary spaces by channeling imperceptible forces that form a new diagrammatic apparatus that is crisscrossed with esoteric potentialities for reimaging self and world from the vantage point of an outside within or a down below. Unlike the One of neoliberal capitalism in which the only meaning that matters is the bottom line, Carrington's practice produces pataphysically pliable multitudes that demand discretion (subterranean sensitives to almost invisible perceptual darknesses) rather than brute calculation. In all cases, studioing is not a preplanned space and time in which individuals experiment with their creative capacities/skills but is rather a space-time experiment in and for creation that invites (a) unprofessional excesses, (b) awkward disorientations, and (c) desirous swerves as evidence of the visitation of dark precursors into the circuitry of creative production and risk management that define the white, cybernetic control of the neoliberal world. Whereas creativity produces value, creation produces only traces of inoperative works that are valueless to those who belong to the cult of neoliberalism. While individual actions of the creative entrepreneurial individual might appear unprofessional, awkward, or mad at times, two facts are guaranteed: Their orientation around the question of profit is always consistent/their madness is always a madness *within* the question of value production (never outside it), and their works are never truly inoperative (some value is always recuperable or extractable). Thus, any awkwardness, unprofessionalism, or madness exhibited by the entrepreneur is merely superficial as neoliberal capitalism will eventually adjust itself accordingly in order to reclaim such behaviors as signs of prodigal genius.

In short, neoliberalism appropriates alchemical creation (that is, how transmutation disrupts division between the possible and the impossible) only to denude it of its critically and radically unsettling force in the name of value production. If capitalism is its own form of sorcery, then it has developed certain infrastructures (including algorithmic rules) of control that enable it to alchemically transmutate drifts of desire into a calculable flux. Opposed to the reduction of cosmological creation to entrepreneurial creativity, research-creation as experimental studioing returns to the alchemical potentialities of what is maximally unprofitable: the vestigial, useless, (non)knowledges of dark precursors that creative entrepreneurs try to wrangle and exploit through their various risk-management calculations yet cannot quite domesticate. The pataphysical alchemy of research-creation can act as a counter to the capitalist sorcery of control, and the space-time of the studio (in all its pliable manifestations) can safeguard an undercommons full of unprofessional, awkward, and mad swerves.

If neoliberal capitalism and its creative economies end up risking nothing and creating nothing, what happens when this *nothing* is absolutized? It is our contention that fascism is an acceleration within capitalism, exposing how the "creative" surplus of capitalism is actually hiding a radical nothingness that is not creative at all but actually potently destructive. While citing various warnings offered up by Deleuze and Guattari throughout this chapter, here we want to pay close attention to their description of fascism as a cancerous Body without Organs. Deleuze and Guattari write that taking the risk of schizzing the self can result in the "cancerous BwO of a fascist inside us," which, on a molar level, results in a paranoid fetishization of "war and money."[34] In the name of war and money, fascism risks everything, ultimately revealing the emptiness and nihilism at the heart of these pursuits. As the seemingly endless creative risk-taking of capitalism expands, it becomes increasingly rigid, hard, and cold, collapsing in on itself. Whereas the curvature of ouroboros repotentializes itself by returning to its origins, capitalism's spiral ends up sacrificing such potentiality on the altar of its own hubris. Both the secularism of neoliberal capitalism and the fanaticism of fascism are appropriations of esoteric knowledges and practices, but without the possibility of risk catalyzing creation.

Missing is the careful calibration between risk and creation in the alchemical studio of research-creation (or, what Maier referred to as the balance of the senses and the intellect in his alchemical book *Atalanta fugiens*). Speaking of such calibration, Deleuze and Guattari write, "Tearing the conscious away from the subject in order to make it a means of exploration, tearing the unconscious away from significance and interpretation in order to make it a veritable production: this is assuredly no more or less difficult than tearing the body away from the organism. Caution is the art common to all three. . . ."[35] Through alchemical studioing, transmutations happen that tear organs from their functional role in organisms, that tear use from predetermined functions, that tear processes from common sense meaning. Darkness emerges from the drift. This process always involves a certain violence: a risk of cancerous overgrowth in which life eats itself or turns against itself. In short, research-creation must take risks if creation is still a possibility, but these are measured risks, informed by Dufourmantelle's notion of "secret intelligence," that emerge in relation to the ways in which rules and drift catalyze certain forces, intensities, speeds, durations, and temperatures within the darkened space-time of the studio.

Here we reach a threshold that can only be crossed via the next gate: the way in which the push and pull between creativity and risk within creative economies of neoliberalism ultimately ends up rigidifying into the manipulative, cold, and hard protofascist personality type (always already present though somewhat masked in the entrepreneur-as-sorcerer of money and war). Perhaps fascism is what happens when neoliberal capitalism *forces* the inoperative to operate (produce value) *or else* destroys it (or at least attempts to through all-out war) according to the calculations of an agitational and accelerated algorithmic empire. The cult of creativity reveals itself to be a cult of death.

Gate Four

The Antifascist Politics of Studioing

The Sound of Fascism's Revenge

We offer a quote from the Afrofuturist cult classic *Space Is the Place* (1974) as a gateway into the question of the antifascist political potentials of alchemical studioing practices. Dressed in a far-out, futuristic, pharaoh's robe, the unmistakably groovy Sun Ra narrates his exploration of a distant planet's alien terrain. He observes that the Earth is dominated by "sounds of guns, anger, frustration" whereas his current, undisclosed planetary location has decisively "different" "vibrations." For us, Sun Ra prompts reflection on the sonic signature of global fascism, which is defined by the booming sounds of militarism but also the more subtle frequencies of affective frustrations, which cumulatively might manifest in outward expressions of hate and rage directed toward the perceived threat of unwelcomed others. But more importantly, Sun Ra also implores us to loosen our sonic attachments in order to hear heterogeneous frequencies and thus acquire tastes for unfamiliar flavors that suggest new forms-of-life, new kinds of bodies, and new relations to reality. Indeed, the movie is, as we will argue later, an expression of the alchemy of research-creation in which the cosmos itself becomes a kind of infinite musical studio and Sun Ra's Arkestra becomes a dark infrastructure for discovering and inventing an antifascist, black pataphysics.

Such an experiment is now more urgent than ever, as we are facing a global upsurge of protofascist,[1] aspirational fascist,[2] and/or late fascist[3] political tendencies emerging from within neoliberal capitalism and its contradictory forces. Indeed, the specter of fascism has returned with a vengeance, both on the fringes of power in the form of ultra-right-wing, reactionary, overtly neo-Nazi terrorist groups and at the centers of power in the form of militant ethno- and religious nationalisms. Upheavals caused by a series of global crises and the weakening of influence of democratic institutions have created conditions conducive to the fomentation of new, global fascist politics in the United States, Hungary, Italy, Poland, Brazil, and other countries. In the most extreme versions, these movements and organizations claim their Aryan heritage is under attack by a cabal of powerful Jewish families they call ZOG (Zionist-occupied government) that secretly controls the culture industry, financial networks, and nation-state governments with the hidden agenda to eradicate the white race. They draw upon a host of symbols from Nazi propaganda, pagan rituals, and Judeo-Christian imagery while valorizing fascist leaders such as Hitler and Mussolini. And they actively pursue acts of homegrown terror. In the United States alone, these groups have engaged in murder, the planned kidnapping of government officials, and attempted sabotage of urban infrastructure, all to "accelerate" the impending race war. Indeed, in 2020 the US Department of Homeland Security ranked domestic, violent extremist movements fueled by racism and nativism as the single most dangerous threat facing the country. In short, fascists at the center and at the periphery are on the move, growing in strength, pushing political and social systems near and dear to the project of liberal democracy to the breaking point.

The question we pose in this gate is as follows: What resources (spatial and temporal) does the pataphysical alchemy of studioing provide to resist the lure of fascism? To answer this question, we will advance a theory of fascist subjectivity and sociality and then argue that the practice of studioing provides a space and time wherein the affective and existential dimensions of fascist subjectivity can be suspended. At first, this might seem like a strange claim to make, as there certainly are fascist artists who undoubtedly work in studio settings. Indeed, authors such as Mark Antliff have demonstrated that a number of French, modernist artists supported the fascist project in the first half of the twentieth century, finding in the avant-garde an aesthetic of revolution that had the power to overthrow existing institutions in the name of a fascist "new man."[4] Since fascism is, in part, predicated on mythmaking, artists had an important role to play in promoting the fascist political agenda. If Antliff is correct, then we must proceed with caution and not overly romanticize the political implications of studioing for any antifascist project. Having said that, we nevertheless find within the porosity, profanation, contamination, and pliability of the concept of studioing (as a space-time matrix) a certain potential that can interrupt and render inoperative the defining features of fascism. Throughout this book we have named this pliability the "alchemy" or "pataphysics" of studioing, and it is precisely the unique features of this occult-like, dark power that are simultaneously captured by fascist appropriation and resistant to such appropriation. The work of this gate is to focus explicitly on the pataphysical alchemy of studioing in order to demonstrate its antifascist implications.

This is not an argument that studioing should be instrumentalized in order to fight fascism. Such an argument would undermine our claim that studioing is, in essence, a vestigial rather than prosthetic practice (see gate five). Indeed, it is our wager that fascism instrumentalizes the experimental qualities of studioing for overt political aims. Simply exploring the potentialities of the space-time of research-creation that emerges in moments of studioing suspends this instrumentalization and as such offers a brief glimpse of an antifascist alternative. This alternative will remain unnamed, manifesting only in a negative (anti-) formulation so as to keep open-ended what kind of collective life can emerge at the intersection of alchemical recipes/patarules and drift. In order to explore the an-archic space-time opened up through studioing, we conclude the chapter with a return to Sun Ra's film *Space Is the Place*, which manifests the suspensive intensities of black, alchemical, and pataphysical research-creation in the space-time of cosmic studioing.

Fascist Subjectivity: An Alchemical Composition

In this section, we offer four brief theories of fascism that, in the end, support one another (despite antagonisms between various perspectives). Instead of a synthesis, we propose an alchemical composition that mixes theories into a clump that does not resolve tensions so much as it produces an effect that transforms how we see fascism. Enabling this alchemical procedure is a fundamental connection between drifting movements and heterogeneous excesses in each theory that, when calibrated through studioing practices, disrupts fascist attempts to harness certain energies, intensities, and forces to consolidate and homogenize power.

Benjamin on Looseness . . .

In a footnote for the second version of the essay "The Work of Art in the Age of Mechanical Reproduction," Benjamin argues that revolutionary class consciousness "transforms the structure of the proletarian masses."[5] While a class might be perceived from the outside as "compact," class consciousness generates a "loosening"[6] of the masses from the inside. Opposed to this we find the "impenetrable, compact entity"[7] of the petty bourgeoisie, which is not even a class for Benjamin. Compactness means that consciousness has no critical distance from the immediacy of everyday life, which is defined by a mechanical, reified conformity to the rituals of capitalist production and consumption. This compact mass is given over to "emotional"[8] reactions with little space or time for thinking or experimenting with forms of becoming that might challenge what is taken to be the status quo. Ominously, Benjamin warns that fascism itself is a mobilization of this compact mass, erasing all space and time for difference in exchange for suffocating proximity and sameness—all of which are synonyms for an underlying reification process of consciousness and perception. Adorno certainly drew upon this footnote in order to later theorize fascist subjectivity as essentially hard and cold (incapable of affective engagement with the world beyond rigidifying agitation).[9] Victor Klemperer recalls a frightful scene of marching witnessed in Nazi Germany:

> I have seen many parades in my life, both in reality and on film;
> I know what Prussian goose-stepping is all about—when we
> were drilled on the Oberwiesenfeld in Munich the command
> was always: You've got to do it at least as well as they do it in
> Berlin! And yet never before, and, what is more, never again,
> have I seen anything to match what I saw that evening, despite
> all of the parades in front of the Führer and all the Nuremberg
> march-pasts. These men kicked their legs so high that the tips
> of their boots seemed to rise up above their noses, all in a

> single sweeping arc, all as one leg, and the posture of all the bodies—no, this one body—was so conclusively taut that the whole movement appeared to freeze, in a way that the faces had already frozen, so that the troop as a whole gave the impression of being utterly lifeless and frenziedly animated at the same time.[10]

The rigid conformity, precision, and exacting movement of the Nazis erases difference, deviation, and the freedom to drift in wayward directions. The result is a single, frozen body that seems to lack life, transforming bodies into a uniform automaton with a single teleological goal in mind: mass murder. This frozen body is equal parts "lifeless" (cold, hard) and "frenzied" (agitated) so that it cannot feel anything beyond its own propulsive acceleration toward its destruction.

The revolutionary consciousness of the proletariat, on the other hand, involves a loosening up of this tight conformity, insisting on the space and time for radical thinking and feeling that orients itself toward that which is farthest away (a not-yet-become secret signal from the future), not what is proximate (the immediacy of blood and soil). In short, loosening makes space and time for associations with others (otherness), and this, for Benjamin, is the definition of political solidarity. Rather than a tightly drawn knot, *solidarity is a loose knot* that allows for both physical movement of bodies and for the movement (drift) of ideas and sensations that would otherwise be controlled by fascist marching orders (and the underlying paranoia that any deviance is dangerous).

Using our terminology, the proletariat is a class that drifts away from conformity of the marching mass to form open-ended, loose social (de)formations. The proletariat is an inoperative mass that prefers not to abide by the laws of the fascist leader and thus does what it is not supposed to do, thinks what it is not supposed to think, feels what it is supposed to be indifferent to feeling. Class warfare is an alchemical experiment in what can be perceived, how bodies connect with one another, and how space and time are generated out of collective movements of drift. It is an experiment in loosening knots so that potentiality is not simply put to work in the name of capitalist profits or fascist imperialism but rather opened up, experimented with. A loose class is antifascist because it is perpetually distracted, looking elsewhere, toward an alternative horizon of possibilities beyond fascist capture and control. The fascist leader cannot hypnotize a loose class, cannot keep it in its place, cannot give it marching orders. Instead, the loose, distracted, inoperative class prefers not to pay attention as it is busy doing something else, somewhere else, falling off line (but also taking up queer lines).

But the looseness is not total dissipation. Looseness is, for Benjamin, the condition for the production of solidarity. Solidarity here is not about forming hard, tight bonds that exclude difference. This would result in massification (and, ultimately, fascism's return). Masses lack solidarity. They are conglomerates of isolated individuals that are cut off from each other (through hardness and coldness). Looseness opens the masses up to the outside, meaning that it allows for some critical distance from the hypnotic powers of the fascist leader while also providing breathing space for bodies to be affected and be affected by other bodies outside of fascist agitation. At the same time, looseness provides a minimal curvature or swerve that clumps bodies together so that dissipation does not accelerate (thus preventing any kind of solidarity from forming). The rules of engagement that emerge from within this loose assembly of bodies are not so much constraints as *permissions* to experiment with an inoperative form-of-life that prefigures something beyond capitalism and beyond fascism (drift as d-rift with dominant political and economic forces structuring social relations under capitalism).

Interestingly, Benjamin argues that "class-conscious proletariat forms a compact mass only from the outside, in the minds of the oppressors."[11] In this sense, the internal emergence of drift is *invisible* to those on the outside looking in. Drift is a movement in and through darkness, which the fascists cannot penetrate with their algorithmic gazes. Class is the incubation of a anticapitalist and antifascist (an)aesthetic, a way of becoming-invisible or dark so as to avoid capture. Looseness of drift is an infrathin difference between the class and its position within the social relations of production—a minor difference that is so discrete that it appears to be no difference at all and yet makes all the difference. One needs to have cultivated a new type of vision that can see in the darkness to apprehend the appearance of a class-consciousness proletariat from inside the mass of workers distributed within a factory. An alchemical transmutation of the senses through new revolutionary organs of perception is called for. Yet to the capitalists and fascists, such organs are merely vestigial, lacking capacity for value production. For this reason, looseness remains invisible . . . until it is too late.

. . . *and Bataille on the Heterogeneous* . . .

For Bataille, fascism is the outcome of the complex relationship between conflicting social (de)formations: the homogenous and the heterogeneous. Of the two, homogeneity is the most fundamental tendency in society, giving sociality a certain structural stability and giving rise to socially normative values. The homogenous society has its base in economic production, and on this level, Bataille does not appear to be radically different from most Marxists. Summarizing the first part of Bataille's thesis, he writes, "Homogenous society is productive society, namely useful society."[12] Usefulness in this context is measured in terms of the productive potential to produce money (profits), and in this sense, usefulness is calculability. The homogenous dimension of society is controlled by those who also control the means of production (the capitalist). The capitalist attempts to maximize the usefulness of a given social product in terms of the calculation of profitable returns. As Bataille argues, "Every useless element is excluded"[13] from such calculations. The useless is largely invisible to the calculous of the capitalist as it does not generate profits. Within this system, the proletariat has an ambiguous role to play. On the one hand they are excluded from profit, but they are included through their labor. When laboring, the proletariat are psychologically integrated into homogeneity, but outside their jobs (and thus outside of white, algorithmic surveillance), the proletariat is a stranger to the homogenous logic of labor and profit. Since homogeneity is predicated on economic productivity, the invisibility of the proletariat as stranger becomes a problem.

Indeed, for Bataille, the homogenous society is always under threat from heterogenous forces in the form of internal disruptions to the production process. The state attempts to ensure stability through processes of adaptation. In other words, the state attempts to neutralize the destabilizing powers of the heterogeneous. Yet this is a difficult if not impossible task because, by definition, the heterogenous escapes scientific calculation. For Bataille, ". . . the object of science is to establish the *homogeneity* of phenomena; that is, in a sense, one of the eminent functions of *homogeneity*" and therefore, "as a rule science cannot know *heterogeneous* elements as such."[14] Science, is an essential dimension of homogeneity and by extension the state. The state functions to assimilate the heterogenous, but, because its framework for viewing social, political, and economic relations is through the lens of scientific inquiry, it cannot ever fulfill its function and is *"deprived of any functional satisfaction."*[15] Whatever constitutes the heterogenous element of society, it has gone by various names, including the sacred (as a mysterious and impersonal force), which has to be separated from the homogenous via taboos. But beyond the sacred, there is an even more generic name for the heterogeneous: *"unproductive* expenditure" that is of no use (and thus does not produce profit).[16] We will go into more detail about this concept in the next gate, but for now, we want to turn to the relationship between this unproductive surplus and fascism.

Fascism and the fascist leader are "incontestably part of heterogenous existence."[17] Democracy (at least the liberal variety) concerns consensus (and thus assimilation) into the homogenous order of society. Fascism turns toward the heterogenous in order to harness the powers, forces, and intensities that place the homogenous under threat of dissolution. The fascist leader's power over followers is likened to that of hypnosis that creates an "affective flow that unites him [the leader] with his followers—which takes the form of a moral identification of the latter with the one whey follow," and this affective identification through hypnosis creates "increasingly violent and excessive energies and powers that accumulate in the person of the leader and through him become widely available."[18] In short, the fascist leader becomes *sacred*, tapping into and channeling the pent-up affective surplus of the heterogeneous elements of society into a mobilized force. But what is most interesting in Bataille's analysis is how the homogenous *in general* needs the heterogenous (in the form of the king, the leader, the law) in order to found itself and to make itself operative. At the heart of the homogenous is a heterogeneous surplus that it is dependent upon yet disavows. This heterogenous element is a sovereign, "imperative agency"[19] that embodies a necessary though disavowed violence, often sadistically directed at those other heterogenous elements that the homogenous society cannot calculate or assimilate (impoverished populations, or those workers who have left the factories and have thus become strangers). In this way, the homogenous and heterogenous dimensions of society are not antithetical but complexly dependent on one another, with the thrust of this dialectic oriented in the "general direction of existing homogeneity, namely, with the interests of the capitalists."[20] Fascism is just an extension of this logic, representing the full integration of the heterogenous into the homogenous domain of the state (the exceptional, heterogenous leader *as* the embodiment of the state, the military, and the economy).

If this is that case, then what can one do in response to total fascist mobilization of the violence of the heterogenous in the name of the homogenous productivity of the state? Infamously, Bataille leans into the heterogeneous. This move has led to many scandals surrounding Bataille. Indeed, Bataille has a long and complex relationship with fascism. While Bataille was a harsh critique of fascism, he has been subsequently accused of his own brand of "left fascism," especially concerning his relationship to secret societies and occult fascination with death, myth, sacrifice, affective revolution, heterogeneous expenditure, and the sacred.[21] Yet these charges miss something important: that Bataille was attempting to attack fascism *from the inside*. Bataille clearly states that "total oppression" is the result of the "unlimited actualization of imperative forms"[22] to force unity and concentration, or total fascist integration of the homogenous and heterogenous (state and sovereignty). Here Bataille's emphasis on the violent force of the sovereign imperative underlying unification and homogenization sounds very much like Benjamin's analysis of fascism as a tightening or closing off of the masses from any contact with difference and otherness. Although a critique of fascism is clear, this does not mean that Bataille critiques the heterogenous as such. In his text titled "Toward Real Revolution," Bataille writes that it is essential to "know how to appropriate the weapons of our adversaries,"[23] meaning the power of the heterogenous that is weaponized by the fascists. In this sense, Bataille is not far off from his friend Benjamin who also neighbored fascism in order to understand it and render it inoperative/vestigial (in order to open up its key concepts to new use beyond fascism). While not relying on the revolutionary potential of the working class, Bataille nevertheless was searching for the "inexhaustible wealth of the forms particular to affective life" that "differ from those already in existence, as different from present or even past communism as fascism is from dynastic claims."[24] Affective forces of the heterogenous have yet to find representation, but in their dark, heterogeneous, useless, and (ae)aesthetic state, they carry a certain amount of subversive potentiality.

Take for instance the headless figure of Acéphale, a key figure for Bataille and his secret society who will be discussed in more detail when we travel through the next gate of this book. For Maurice Blanchot, the headless nature of Bataille's collective ritual practices contrasts with the "military or fascist group where each member of the group relinquishes his freedom or even his consciousness to a Head incarnating it."[25] The acéphalic body lacks the tightness of the concentrated masses, but most importantly, it does not cut off its own head in order to simply rely on the head of the fascist leader. Instead, it is a body that is adrift, unable to be unified under a centralized agency. It is a loosely structured body that is without center and that is full of intensities of heterogenous affects. The fascists might sacrifice individual heads, but they do not risk a state of headlessness for long. Instead, the head of the leader reconstitutes the fascist body, and as such, fascism has a secret investment in the homogenous, turning the heterogenous into a monstrous behemoth guided by the leader. The transgressive power of (headless) heterogeneous expenditure is harnessed in the name of repressive, nationalist homogeneity (the head of state). Against this, Bataille turned to the heterogenous possibilities of the heterogenous, in whose darkness foments new affective subversions (rather than fascist imperatives).

. . . and Deleuze and Guattari on the Drift of Desire . . .

In the preface to *Anti-Oedipus: Capitalism and Schizophrenia* by Deleuze and Guattari, Michel Foucault clearly states that the "strategic adversary" of the book is "fascism"[26] not in the sense of political movements, but in terms of subtle, everyday forms of fascism that might manifest through the love of power and hierarchy as well as a desire for one's own subordination and domination. While Benjamin and Bataille focus on large-scale social and political movements, Deleuze and Guattari worry that fascism could take hold of an individual's everyday psychological dispositions and that even "liberal" individuals might come to *desire* fascism. Such desire operates below the level of subjective beliefs or even psychological states. As such, the techniques of schizoanalysis that Deleuze and Guattari offer promote a "non-fascist life"[27] that must radicalize desire, revolutionize the ways in which it structures society around fascist social forms.

For Benjamin, fascism emerges from a dialectic between tightening and loosening of masses, and for Bataille, fascism emerges from a co-optation of the heterogeneity by homogeneity. For Deleuze and Guattari, fascism is located in the oscillation between sedentary (territorialized) desire and desirous drift (deterritorialization). Desire, for Deleuze and Guattari is, as stated in the first gate, an impersonal, dark infrastructure that both supports and pushes up against the limits of any given social, political, or economic assemblage. On the one hand, capitalism deterritorializes flows of desire: flows of money and circulation of commodities, flows of workers and information. Capitalism "schizophrenizes" society.[28] The dynamism of capitalist modes of production and consumption *loosens up* stratified social bonds and overcomes territorial boundaries (of nation states) so as to maximize flows through deregulation. On the other hand, capitalism places certain limits to the productivity and polyvocality of desires that it unleashes. Deleuze and Guattari summarize: "The more the capitalist machines deterritorializes, decoding and axiomatizing flows in order to extract surplus value from them, the more its ancillary apparatuses, such as government bureaucracies and the forces of law and order, do their utmost to reterritorialize, absorbing in the process a larger and larger share of surplus value."[29] In other words, capitalism develops the capacity to absorb heterogeneous desires and make them operative under its axioms, and thus reterritorializes the flows it catalyzes, via the family, the nation, and the state as spaces of capture and control. This tendency of capitalism to reterritorialize flows culminates in the fascist state, which is "capitalism's most fantastic attempt at economic and political reterritorialization"[30] insofar as there is a perfect merger of industry, war, and state. As Bataille would argue, there is a complex relationship here between the heterogenous pluralization of desires within capitalism and the subsequent attempts to homogenous such pluralism within a state (now surcharged by its encounters with excess).

On a subjective level, the self becomes calcified into a potentially "reactionary" and "paranoic" "fascist tendency,"[31] which culminates in the formulation: "I am one of your kind, from the same place as you, I am a pure Aryan, of a superior race for all time."[32] As one can easily see from this formulation, the fascist tendency is to create strong identity politics predicated on clear, hierarchical boundaries between self and other. If schizophrenia is in some sense unleashed within and through capitalism, when a speed limit is placed upon it, schizophrenia transforms into paranoia (in which there is one source of meaning and all desire is coded according to a despotic power). The paranoid subject is defined by "identification of the individual with the group, the group with the leader, and the leader with the group"[33] with the goal of getting closer to the center of power, never being on the periphery (except out of necessity to protect the center). Coldness and hardness set in, or what Deleuze and Guattari refer to as fascist "sedentary" desire[34] in which desire attaches itself to power and the worship of strong leaders/essentialized racial categories/mythologized homelands. The productive capacity of desire as dark infrastructure capable of loosening boundaries, borders, and territories is replaced by the constricting power of the "herd instinct,"[35] which controls the movements of desire according to global aggregates (such as "the people" or "the nation" or "blood and soil").

How to respond to the increasingly paranoid, reactionary, and rigidifying nature of capitalist territorialization of desire? Schizoanalysis is an attempt to break free from fascist hardness and coldness in the name of "nomadic and polyvocal"[36] desiring lines. It is a procedure for rupturing fascist libidinal interests into a herd (or what Benjamin might refer to as a mass) by unleashing desire that has the capacity to disorganize the striated, fascist social field that is organized hierarchically and schematically. Deleuze and Guattari summarize the productive nature of schizoanalysis as unleashing "pure positive multiplicities where everything is possible, without exclusiveness or negation, syntheses operating without a plan, where the connections are transverse, the disjunctions included, the conjunctions polyvocal, indifferent to their underlying support, since this matter that serves them precisely as a support receives no specificity from any structural or personal unity . . ."[37] Notice how schizoanalysis substitutes rigidity/tightness for pliability/looseness (a multiplicitous, polyvocal, polymorphous, transmutated self).

Guattari[38] argues that fascism is ultimately about keeping desire "in line" whereas schizoanalysis is about setting desire adrift, which is an alternative space-time below rigidification, manipulation, and instrumentalization—a nomadic or rhizomatic space of becoming. Importantly, he writes, "At home, a child 'off the track' is put down, and this continues in school, in the barracks, in the factory, in the trade union, and in the party cell. You must always stay 'on the right track' and 'in line.' But by virtue of its very nature, desire always has a tendency to 'stray from the subject,' to 'get off the track,' and to drift from its proper course."[39] Fascism takes hold on a child by keeping desire on a certain track. Here, we want to place added emphasis on the distinction between flux and drift introduced earlier in the book. Capitalism territorializes desire into calculable flux (which can operate under capitalist algorithms), but desire has another s-pacing, drift, which opens up to improper courses, swerves, and queer lines of flight. It is the drift of desire that is a dark infrastructure that enables Benjaminian looseness through contact with Bataille's excess. Although Guattari[40] warns that everybody wants to be a fascist, meaning that the *flux* of desire can and does plug into paranoid circuits, at the same time, the *drift* of desire (as another s-pacing) is always already turning away, toward the outside. Thus, drift is the key to an antifascist schizoanalysis, which essentially intensifies and extends the drift in order to create new pleasures and new flavors of life that push up against the dulling sensibilities of coldness and hardness (homogenization, rigidity, and tightness of paranoia) found in fascism.

In this sense, Deleuze and Guattari do not retreat from the drift of desire unleashed (deterritorialized) by capitalism but rather embrace such drift, lifting the speed limit that culminates in fascist paranoia. They write, "For perhaps the flows are not yet deterritorialized enough, not decoded enough, from the viewpoint of a theory and a practice of a highly schizophrenic character."[41] In other words, to live an antifascist life one has to experiment with decoding the drift of desire within its flux. The drift is the imperceptible and infrathin becoming within flux, which is a flow that is amenable to capture and measure. But Deleuze and Guattari are far from merely celebrating unchecked accelerationism of drift beyond the speed limit of fascism. They are clear that such experimentation is not always positive and productive. Proliferation of lines of drift can lead to increased freedom and creation of new bodies, new mental acuities, and new modes of antifascist life but it can also produce cancerous formations. This is especially true when the deterritorialization is done without a certain level of caution. The speed limits of capitalism cannot be broken too quickly or all at once, as we have already discussed in the third gate.

In short, capitalism deterritorializes desire and then reterritorializes it in the form of fascism. Deleuze and Guattari argue for an ethic of desire in which deterritorialization is valorized but not too much of it and not too fast. Indeed, pure, deterritorialized accelerationism is as destructive as capitalist reterritorialization. For Benjamin, antifascism concerns the question of solidarity in looseness; for Bataille, it concerns the question of enjoyment of the heterogeneous without imperative force; and for Deleuze and Guattari, it concerns the cautious deterritorialization of desire. All put certain breakers on that which they perceive as antifascist (de)formations: looseness, heterogenous excess, and desirous deterritorialization. In other words, there is a dialectic of drift (as noun and verb) and rules at stake in all these models that prevents either calcified masses (marching orders) or accelerated flux (self-destructive whizzing).

. . . and Moten on Blackness.

On our reading, Moten's paraontology of blackness not only enables us to clump together the key themes of the previous authors but also enables us to more directly locate antifascism as an inherent feature of blackness.[42] At the same time, Moten also demonstrates how blackness is precisely the radically heterogeneous element that must be perpetually subjected to authority of a white, heteronormative, neurotypical, homogenizing law, otherwise its very presence threatens disruption. And finally, more than any of the previous authors, Moten enables us to theorize the space-time of the studio as an antifascist space. Spatial and temporal considerations have played out in almost all of the antifascist theories outlined above. For Benjamin, antifascist forms of association appear in the space of the general strike, spilling out into the space of the street, producing a loose solidarity between workers, students, and citizens. For Bataille, ritual practices at night and often in the woods (as outlined in the next gate) offer the space for experimenting with heterogenous affects. For Deleuze and Guattari, the flat space of the nomad enables desire to drift off-line. But what we find most compelling about Moten's work is its clear articulation of studioing as the space-time for the antifascist alchemy of research-creation.

Moten refers to blackness as an "ontology of disorder, an ontology of dehiscence, a para-ontology"[43] that manifests itself in the "incalculable rhythm of the life of things."[44] To unpack this pithy formulation, we want to highlight the "paraontological" status of blackness. This means that blackness is somehow outside of ontological claims. It might very well precede ontological speculation, but for this very reason, it itself cannot be located within an ontological framework or achieve the status of a "first principle" of philosophies of being. Blackness is what is presupposed by ontology while also escaping ontological definition. Second, if this is an ontology, it is an ontology of "dehiscence," or a loosening of ontological essences. On a political, rather than philosophical, level, blackness would thus lack political being, and instead be relegated to the "undifferentiated"[45] field of the social. For us, the undifferentiated social is precisely the loosening necessary for new solidarities to form through contamination that would not be allowed within the tightly bound (protofascist) masses.

From the paraontological perspective, desire shifts from desire for the standard language (of the homogenous) to desire for linguistic experimentation and invention (of the heterogeneous). Moten writes of the loss of language as a lament for a home and shared spatio-temporal framework for understanding one's self and one's existence, but also of an emergent desire for a "shifting schemata of a fleshy historicity in which language moves to connect a vast, differential range of unmoored unowning."[46] This desire for the unmoored is a desire that drifts from learning the standard in wayward directions. It drifts off-line toward a new speech, toward linguistic experimentation that cannot be reduced to merely degraded forms of standard speaking. The desire of speech is, in other words, deterritorialized from the territorial bounds of the standard (and thus of deviancy and parody of what is "appropriate" locution) in order to generate paralinguistic, profane utterances of paraontological nonbeing.

The "incalculable" quality of blackness means that it will forever elude the empire of algorithms. It will remain invisible or (an)aesthetic to such measures and thus be a perpetual stranger to what Bataille refers to as the homogenous society of productivity and profit (ruled by capitalists). Blackness is radically poor, a minimal state, or infrathin difference between nothing and everything. This means that black people cannot even "own" blackness as a property (as an ontological marker of being). Radical poverty is the result of radical dispossession (economic, political, and even ontological). But for Moten, this is not merely a state of bare life and brute survival to be overcome through an expression of sovereign power (and here the ambivalence of sovereignty in Moten's thought runs up against Bataille in interesting ways), but also reveals a minimal potentiality for what he calls a "general commonness and openness of *a life*" that is not the standpoint of a particular individual or race but rather of immanent social life, which offers solidarity and affiliation with those "who have nothing and who, in having nothing, have everything."[47] The nothing of blackness is therefore never empty or void, but rather an openness (an ouroborotic twist). In sum, the undifferentiated, heterogeneous, incalculable, and impossible position of blackness within a world dominated by and through white algorithmic calculation is not necessarily a state to be negated (in order to achieve political recognition/a standpoint). Instead, one can choose to stay in this "undercommons, underground [crypt]"[48] in order to study blackness and to celebrate its destituent potential for another kind of life.

Here Moten gestures toward a spatial and temporal theory of the studio as the space and time of black, fugitive study. This is a space-time characterized as a "mobility of place" and as a "fugitive field of unowning."[49] Stated differently, the space-time of study (what we call studioing) is constantly mobile, pliable and flexible, incapable of being fixed on any map or territory. It is also fugitive, meaning it pirates existing spaces for resources. And finally, it is not about ownership or property. Instead of property, studioing concerns free *use* and is thus open to anyone. It therefore partakes of and directly adds to the undercommons, or a commons that is cryptic, and thus, is a studious secret (a secret that nevertheless gives itself away as part of the commons). Adapting Moten's formulation, we would argue that studioing is an initiation into "the social materiality of no place, of Having No Place, as a place for study."[50] The activity of study in this fugitive field is described in a-disciplinary terms as an "aesthetic sociology or a social poetics of nothingness."[51] It is, in other words, a space and time for returning to the "nothing" that lies between art and science, pleasure and knowledge, hand and head. This nothing—the gap that is so minimal that it cannot even be named with any accuracy and thus becomes a point of indiscretionary contact—is actually full of unexplored new flavors of life that can begin to take shape through studioing in the undercommons and the production of paralinguistic experiments in living together.

The practice of studioing is not merely a-disciplinary, but is also an occult practice. Moten gestures in this direction by naming study as a "mystical" experience, or an experience of somehow encountering the excesses of the outside (the nothing of blackness). But perhaps more importantly, Moten describes the disavowed or fallen poetics of the undercommons as an "alchemy of nothingness."[52] An alchemy of nothingness is an experiment that emerges in moments of radical dispossession. Instead of a translation of languages, it is really a transmutation of language itself, meaning a reorientation of language away from the productive and effective speech act toward the inoperative or "infelicitous speech act" that "fails, ultimately, to achieve an intention."[53] Again, this failure is not to be mourned but rather celebrated as an opening up of a new kind of language that says Nothing (and thus speaks the paraontology of blackness). This is, as discussed in the antechamber of this book, a turn toward the dark tongue of the riddle, which is equal parts esoteric and exoteric, offering an opaque formulation that nevertheless performs its own solution. For Moten, the noncommunicative opacity of the dark tongue of social poetics in the studio communicates the noncommunicative nature of blackness by not communicating within a communication (an inoperative work for sure!).

We can push Moten's gesture toward the primacy of black paraontology and alchemy even further by placing it within our crypto-genealogy. Alchemy's etymology leads us to the ancient Coptic name for Egypt and to blackness: *Khemia* and the black, fertile soil of Nile Delta. When this region was conquered in the seventh century by the Arabic world, Egyptian knowledge of the magical properties of alloys was adopted as *al-Kimiya*. At the very heart of alchemy is a paraontology of black fertility that is (an)aesthetic (unrepresentable and thus uncapturable by white cybernetics). Such blackness can only be indirectly apprehended through the conjuring of dark precursors in the alchemical and pataphysical studio. Furthermore, in Roger Bacon's *Radix Mundi* from the thirteenth century (a text that was itself in dialogue with even earlier Greek and Islamic sources), Bacon posits the *priority* of blackness in the transmutation process leading to the philosopher's stone. In his description of the signs of the philosopher's stone, Bacon writes, "The blackness the Philosophers called the first *Conjunction*; for then the male and female are joined together, and it is the sign of perfect mixtion."[54] Blackness is a sign of mixing that crosses male and female to produce a hermaphroditic offspring (a becoming rather than a being). Thus, blackness in alchemy transgresses boundaries and results in exceptional progeny that drift in the open space between the sexes—blackness is queer, falls off-line, refuses to keep genders on track. Blackness is a return to the amorphous, the chaotic, fluid state of matter (primordial drift). Certain alchemists referred to this state as a regression to a pre-natal state or a *regressus ad uterum* and thus as equal parts a death and a virtual potentiality for rebirth. The subsequent production of the philosopher's stone is described by Bacon as "the Education of the Whiteness out of the Blackness or Black Matter."[55] Whiteness is educated by black matter as the most basic and significant teacher. Whiteness is therefore derivative of blackness, and should remember that it is a *student* (rather than a fascist dictator!). Or in ancient Iranian Sufism, black light is the light of all lights or that which makes visibility visible but which is in itself invisible.[56] Corresponding to this point, blackness, as Moten argues, is a paraontology of destituent potentiality that is the fecund source for all drift. Whiteness is a *reaction* to the teachings of this darkness. If the quintessential image of fascism is the Aryan man emerging from the sun,[57] then the defining image of antifascism is a fugitive escape into the blackness of the alchemical and pataphysical studio's undercommons.

Finally, Harney and Moten situate the space-time of social poetics (or what we call research-creation) in the studio with antifascism. Fascism is, on their view, anti-blackness, pure and simple. It tightens identity around a shared essence (whiteness, for instance); it subjects the heterogeneous to the regime of the homogenous through this tightening; it territorializes desires (or accelerates their deterritorialization toward destructive annihilation) according to algorithms of identity, which attempt to measure racial differences; and it denies blackness, transforming a nothing into a something that can be calculated (a commodified, policed, medicalized flux of populations). Against biochemical inscriptions of blackness (as degenerate or pathological), Moten argues for an antifascist experiment with "bio(al)chemical"[58] production of new paralinguistic utterances (alchemical recipes and spells). The antifascism of studioing concerns a sharing, but rather than a shared identity, it is the sharing of nothing, or, as Harney and Moten call it, "shared incompletion" and thus the "inoperativity of the individual."[59] In this sense, paranoia becomes paralinguistic experimentation, as there is no identity to protect and guard against invasion from a contaminated outside. Much like Benjamin, the authors see fascism as a hardening of individuals into a mass whereas studioing is antifascist precisely insofar as it opens individuals and their communities up to the nothing of blackness, which is a precondition for an antifascist education in the loose clumping of solidarity. Such nothing is useless to fascism, or perhaps better yet, it is invisible, truly fugitive, to fascist algorithmic surveillance and capture.

Emplotment is a fascist form of territorialization that denies the drift toward the excess of Nothing that characterizes drift of desire. Opposed to this, Harney and Moten gesture toward studioing as a space time that "signifies a movement of and in place, a radical and irreducible movement that constitutes our undercommon indigeneity, our shared, native, ante-natal turning and returning."[60] The movement of turning and returning calls up ouroboros, who repotentializes itself through a turning that is also a returning—a loose knot that provides minimal (dark) infrastructure for the formation of solidarity in difference rather than in identity. The time-space of the studio is, in other words, the "bending of time" and a "displaced place"[61] that lends itself to antifascist forms of research-creation concerning new forms of solidarity and useless expenditures out of twisting, looping, and knotting drift.

The Pataphysical Alchemy of Studioing as Antifascist Experimentation

In this section, we want to pick up where Moten and Harney leave off, further specifying the antifascist potentials of studioing through an in-depth analysis of its pataphysical alchemy. In this sense, we want to highlight the aesthetic experimentation of research-creation in the studio as an antifascist practice. Already, Harney and Moten refer to the space-time of study (or studioing) as a "surrealist spot."[62] Perhaps even more interesting is their concluding reference to Sun Ra's and Funkadelic's spaceships as places of study. For Harney and Moten, these vehicles are vessels for antifascist living in the undercommons—equal parts surrealism and science fiction. We want to take up this gesture toward surrealism but swerve it in a new direction. For us, the alchemy of the studio is just as much surrealist as it is sur-rationalist, thus demanding a more rigorous engagement with pataphysical experimentation in terms of what can and cannot be sensed (thus operating on the margins between aesthetic figuration and [an] aesthetic distortion). In the pataphysical alchemy of studioing, fascist solidity loosens its tight grip, fascist territorialization opens up to drift, fascist homogenization gives way to the unnamed and useless affects of the heterogeneous, and fascist whiteness can no longer keep track of the fugitive, queer lines of blackness. As covered earlier in this book, pataphysical alchemy can be defined in terms of these paradoxical dimensions: a science of particulars, an examination of the laws of exceptions, and the production of imaginary solutions. In this section, we will pay particular attention to how these three dimensions embody the antifascist potentiality of studioing to combat fascist tendencies, and in this sense, extend and intensify the drift of Harney and Moten's thesis.

First, pataphysics is a science of particulars. Whereas modern science concerns ordering specifics in relation to larger, more general categories or species (homogenizing the heterogenous, transforming blackness into biochemical identity markers), pataphysics resists these kinds of inductive and deductive maneuvers, opting instead to focus on the production of particulars and their horizontal, analogical contact points.[63] In relation to questions of risk, pataphysics insists on generating particulars that escape from the molar identities of groups. This does not mean that pataphysics retreats from the social field, but rather it invests in social groups that are not recognizable by the state (minor, heterogeneous peoples to come, loose classes that are undifferentiated, invisible, and incalculable). It generates particulars that cannot be labeled in advance as *belonging* to this or that "type" of group. If science classifies, then the science of particulars declassifies, sets adrift. These are singular group formations that do not yet have a name, do not yet fit nicely into the order of things, and thus mobilize desire in ways that overcome limits imposed by and through homogenizing forces, institutions, and practices. Particulars can pole-vault out of their "proper" historical or social context to create strange conjunctions with other anachronistic elements, forces, and styles. Deleuze and Guattari refer to singularities as perpetually in movement, or accumulating along lines of drift characterized by constant conjunction "and . . . and . . . and."[64] In Jarry's pataphysical science, particulars are roaming clinamina within what he refers to a "ethernity" or cosmic energy field that enables them to travel across space and time and produce creative (if not unholy) couplings or loose, diagrammatic knots. For us, ethernity is precisely the generative nothingness of blackness.

Fascist, paranoid, stereotypical thinking is, like modern, capitalist science, predicated on inductive or deductive reasoning. It moves vertically, either shuttling from the general to the particular or from the particular to the general. Such vertical movement often lends itself to hierarchization, which can take theological (as in the Great Chain of Being), evolutionary (in which certain organisms are privileged as more "intelligent" or "sophisticated" than others), and political (in which races are ranked according to perceived superior traits or purified bloodlines) guises. Yet pataphysical experimentation is analogical, meaning that it proceeds horizontally by clumping together seemingly unrelated particulars (. . . and . . . and . . . and . . .) to form accidental becomings. This approach does not erase particulars, but rather explores combinations/becomings, creating diagrams of how particulars interact in novel contexts. It therefore resists the kinds of rigid hierarchical stereotypes that dominate fascism. Against this kind of stereotypical thinking, a pataphysical science of particulars is a much-needed experiment in non-reductive perceiving and thinking that is a-geneological, horizontal, and plays and proliferates heterogeneous series.

Furthermore, pataphysics concerns itself with the laws of exceptions. This might very well appear to be paradoxical as exceptions are exceptional precisely because they do not abide by laws. Yet as the work of Agamben demonstrates,[65] exceptions do not abolish the law, rather they are states that protect the law precisely by suspending it. Such states can be witnessed in any number of historical moments, as when governments proclaim a state of emergency. For us, what is most important is the connection between states of exception and fascism. Indeed, the rise of Nazism to power in Germany in the twentieth century was defined by an extended, indeterminate state of exception, which granted Hitler, as sovereign, absolute power. Or we can think of the (quasi-fascist) state of emergency induced by American-style liberal democracy in the wake of 9/11 that allowed for any number of otherwise illegal incursions into private lives of citizens in the name of safety. Or we can think of the dystopian speculative fiction of accelerationist fascists who revel in total capitalist deregulation as its own kind of state of exception. If this is the case, then what are we to make of the antifascist potentials of pataphysics? Would not pataphysics itself be a proto- (rather than anti-) fascist aesthetic? Here we return to opening worries concerning the connections between avant-garde artists and fascism.

We can resolve this question by turning to Benjamin, who once wrote that if we live in a perpetual state of exception brought about by sovereign decisions *from above*, then it is "our task to bring about a real state of emergency, and this will improve our position in the struggle against Fascism."[66] For Benjamin, the real state of emergency/exception is not from above but from *below* in the form of worker's (and children's) strikes and other revolutionary gestures by the oppressed, marginalized, and excluded in which the solidity of the mass (and its paranoid, stereotypical thinking) is loosened and opened up for new associations (new analogical relations between particulars). Thus, the source of the exception shifts from those in power to those who have little power (institutionally speaking). But there is another important shift here worth bringing to the fore. Fascism always acts to *absolutize* the state of exception, or make it permanent in the name of power. It desires to produce a state (homogenous) out of the state of exception (heterogenous), and thus subject anyone to the possibility of unjustified and unsanctioned violence and arbitrary decisions by a sovereign. Pataphysics insists instead on the *evental* nature of the emergency. Exceptions are events that rupture social formations and organizations, opening up possibilities for new kinds of social groups and subjects to emerge, yet such moments are precisely that: moments or happenings. Benjamin's examples are illuminating in this respect, as both worker's and children's strikes are specific happenings that do not attempt to absolutize the emergency, instead these are events that rupture the rupture of the state of exception (from above). We can thus think of the strikes as moments in which subjects can study the laws of the exception from below so as to unleash the flows of desires and affective intensities otherwise restrained by molar aggregates from above.

And this brings us to another distinction. The state of exception from above transforms everyone into bare life[67] or life that does not have the safety of rights or laws to protect it from sovereign violence. Such an exception reduces life to that which can be instrumentalized and manipulated so as to fulfill the function of a sacrifice. This holds true for historical and present variants of fascism: both of which are politics of sacrifice, death, and destruction. But the exceptional event from below does not insist on such instrumentalization of life. Instead, it is a momentary manifestation of the powers of life to demand something else, to prefer not to abide by the sovereign decision over and against it, and thus embrace its own insurgent powers of (anarchic) and collective, useless expenditure (useless, that is, to regimes of fascism).

As Gavin Grindon argues, although Marx used the phrase "alchemist of the revolution" as a negative critique of revolutionaries lacking a true materialist understanding of class struggle, Bataille might have found this phrase to be a complement insofar as alchemy is the science of particulars and exceptions rather than economic or fascist rules (both of which are *headstrong* and *algorithmic* calculations of money and power).[68] What Benjamin and Bataille are both arguing for is an embrace of the *exception to the exception* (the real state of exception) or the accursed share *that is heterogenous to the fascist appropriation of the heterogenous*. In both cases, the alchemy of pataphysics pushes us into the very motor driving fascism in order to subvert it from the inside—a dangerous maneuver for sure.

Third, pataphysics offers imaginary solutions. Again, we appear to be dealing with an irreconcilable paradox as Jarry is asking us to accept the material reality of "solutions" that are nevertheless "imaginary" and thus nonfunctional (and therefore not really solutions at all). Compare this with fascism's "final solution." In fascism, there is a search for a solution that will end all problems. For such a solution to exist, it must posit a definable problem that is clear, simple, and can be directly solved. The figure of the Jew in Nazism fulfilled this function—the cause of all problems—and the final solution was a mechanism for addressing this cause once and for all. The teleological outcome of this final solution would then be the "eternal" establishment of the Third Reich. Or, one can think of the function of "illegal immigrants" in Trump's fascist line of thinking as a roadblock to "making American great again": these "bad hombres" were the ultimate cause of all American problems and thus erecting a wall (as a final solution) seemed feasible (if not reductive and overly simplistic). Such a solution promises to cure perceived societal ailments, restoring the nation's health. In a recent fundraising email, Trump proposes that "Our Country is being poisoned with the millions of people that are illegally flowing through our borders," and "Our Country is dying from within and nobody is doing anything to stop it." The erection of the wall becomes a central mechanism for ensuring purity. In such cases, problems have to be solved absolutely, without remainder, in order to fulfill a mythic regeneration of society. To do so means transforming complex social, political, economic, and even aesthetic issues into simply stereotypes (what we might today call conspiracy theories) that can be solved in one bold stroke (if there is a leader with such power and vision willing to "speak the truth" and take decisive action). Historical, open-ended becoming transforms into a fixed historical narrative with a predetermined, preordained conclusion.

Or, perhaps even more ominously, contemporary, accelerationist fascisms seem to reject any solution at all, opting instead for suicidal nihilism. All solutions are pathetic attempts to hold on to some notion of the human that has been superseded by capitalist and technological explosions. Fully embracing capitalist speed without limits, there is a telos at work in accelerationism, one that inevitably leads to a total collapse in a moment of fanged jouissance. The warnings posted by Deleuze and Guattari are simply laughed off as reactionary attempts to undermine the power of desire that Deleuze and Guattari supposedly champion. For instance, Land would simply argue that capitalism is already a black hole that one cannot escape, and that any attempt to cautiously organize alternative forms of life immediately can be co-opted by capitalism and are thus part of the problem rather than a solution. The only option is to embrace the nihilism of capitalism full throttle and accelerate it toward a fascist horizon.

Yet, Jarry's notion of "imaginary solutions" suggests something radically different than either of these options. Instead of a final solution (a solution to end the need for further problem-solving), the imaginary dimension of these solutions means that *there is never one final solution*. Instead, what is asked of us is to play with permutations and combinatorials of imaginary solutions, to experiment with variations, to engage with becomings as *so many inoperative and useless solutions*. Thus, the problem never ceases to be a concern, and instead acts as a generator for more variations. This is not some kind of "infinite deferral" of action. Instead, it is a way in which we can encounter potentiality as such: the potentiality for new uses within inoperative works that resists simple instrumentalization. One can think here of Deleuze's analysis of Beckett's TV plays[69] in which figures trace out various repeated patterns on a square, seemingly without end, never "solving" the aesthetic problem of the quad definitively. Indeed, Beckett's plays are complex knots that diagrammatically materialize ouroborotic circuitry that potentializes by turning in on itself. The imaginary *repotentializes* the solution, multiplying it infinitely. In a certain way, we can thus think of imaginary solutions as resisting the fascist tendency to *instrumentalize* solutions (meaning that all solutions have to produce pragmatic, measurable, quantifiable outcomes). Instead of instrumental solutions we arrive at a point of *aesthetic* solutions that never seem to actualize their own powers to "solve" anything so much as generate more imaginary solutions in the name of continued, open-ended experimentation in use. And in terms of contemporary, accelerationist fascisms, the notion of imaginary solutions is radically different from *no solution at all*. The latter is nihilistic to the extreme, simply accepting a certain telos within capitalism toward self-destruction, whereas imaginary solutions embrace the potentials that exist now for producing inoperative works that do not accelerate but rather slow down (hastily, of course) in order to contemplate, plan, and use the world (rather than annihilate whatever remains of it). Accelerationism does not have time for such deviations from its meeting with doomsday. The pataphysical and alchemical gamble seems to be that in and through slow haste in which impossible solutions are continually generated, doomsday will not simply be postponed but abandoned as a destination entirely.

The pataphysical alchemy of studioing offers the dark infrastructure necessary to let loose, to tap into heterogenous forces, powers, and intensities, to allow desire to drift, and to return to the generativity of blackness to produce nothing (a pataphysical formulation if there ever was one) without necessarily capitulating to the desire for fascism. Key here is the *space* of studioing as a space of experimentation with particulars, laws of exceptions, and impossible solutions. This is a space that is supported by and generates dark infrastructures that are hospitable to the heterogenous without toppling back into tight, territorialized, homogenizing spaces. The space of the studio is open to dark precursors, which can effect radical transformations in the composition of bodies, affects, and objects (d-rifting where bodies are found, what they can sense, and what they can produce). This is a space of risk in which tight knots are loosened just enough to let something happen that is unexpected, unanticipated, and for these reasons, threatening to fascist social and subjective formations. It is a space that is loose but not too loose, that enables drift both to move and clump, that embraces the heterogenous but in a way that is supported by infrastructure necessary to avoid collapse into the abyss, and is fugitive yet common so blackness can generate new forms of sociality through the celebration of dark tongues.

In the initiation to this book, we characterize studios in terms of differentiation (porosity), deactivation of authority (profanation), disorientation of inside/outside dichotomies (contamination), and a-disciplinary experimentation (pliability). Now, with references to the antifascist thinkers outlined in the previous section, we can fill out the specific antifascist dimensions of these qualities. Benjamin allows us to appreciate how porosity is a loosening of affective and cognitive ties to hard and cold social and institutional formations indicative of fascist collectivities. Bataille reminds us that the heterogenous can take on anti-authoritarian forms beyond imperatives of force. Deleuze and Guattari pinpoint the antifascist potential in the ways that desire contaminates, defying rigid boundaries, producing analogical affiliations across differences that fascism would otherwise like to keep separate by imposing state and economic speed limits. And finally, Moten finds alchemical, a-disciplinary studioing within the fecund potential of a destituent blackness that constantly presses up against the limits of algorithmic calculability and its logic of functionality. When clumped together, these features identify studios as quintessential heterogenous spaces, or loose spaces of study and research-creation that contain within them distinct antifascist potentiality.

The *time* of studioing is also important to note. Recall that the pataphysics of studioing concerns the paradox of slow haste. The capitalist speed limit might be broken by studioing, but this is not pure accelerationism or whizzing uncontrollably outward. Pure accelerationism leads to death, madness, or a new type of fascism (as opposed to the traditional marching orders of historical fascism). Instead, studioing concerns s-pacing. It is a space but also a certain pace in which the subject subjects themselves to experimentation (risks that are awkward and unprofessional) yet with a certain degree of caution or hesitation that lives in the state of exception between slowness and haste. Such s-pacing is knotted together with the rules of the studio. Perhaps we can even argue that pure speed (accelerationism) is the result of financial capitalism's empire of algorithms, which offer thin rules ruled by the instantaneousness of the calculation over and above any discretion, which calls for more time to think, feel, and perceive differently. The alchemical recipes and patarules of the studio exist precisely in a state of exception between algorithmic rules and discretionary rules, and as such can never be reduced to a thin rule of pure algorithm calculation, and by extension do not induce the whizzing movement necessary for accelerated fascism. In this sense, it is important to remember that Deleuze and Guattari themselves describe the movement of becoming with the following:

"Be quick, even when standing still!"[70] This demand should be read as an ethical injunction. It does not encourage pure accelerationism, but a careful s-pacing. One occupies a space (standing still), but stillness is already intensified and innervated with a quickening. Likewise, the quickening is stilled by standing. There is a slow haste and a hasty slowness at stake that comes about through the discretionary play of rules interacting with emergent becomings. On our read, the s-pacing of studioing practices is an essential dimension preventing either a collapse back into historic fascism (where desire slows and stagnates into marching) or acceleration forward (where the pure speed of desire is fetishized in the form of whizzing).

This argument means that we disagree with Steve McCaffery's thesis in the chapter "The 'Pataphysics of Auschwitz" from his book *The Darkness of the Present*. In this chapter, McCaffery attempts to "expose a 'pataphysical dimension within a part of the internment camp's practical execution that proffers somber consequences."[71] And in the carefully schematized plan for the Auschwitz death camp, McCaffery hears echoes of Père Ubu's bleak observation that "We shall not have succeeded in demolishing everything unless we demolish the ruins as well. But the only way I can see of doing that is to use them to put up a lot of fine, well-designed buildings."[72] For McCaffery, this syzygial contact between pataphysics and the internment camp offers a sobering lesson for those who would celebrate the exception *without exception*. Yet McCaffery's analysis is inadequate in several respects and misses precisely what is antifascist in pataphysical studioing. As we discussed earlier, for Benjamin, there are states of exception from above (Ubu) and states of exception from below (the desire of a minor people, or a people yet to come). This might, in the end, be an *infrathin* distinction within the exception itself, but it is an important one to keep in mind as it both keeps in peripheral vision McCaffery's concerns while also foregrounding the antifascist potential of the very same pataphysical state of excess/expenditure/exception. The pataphysical alchemy of studioing conjures up clinamina, syzygial combinatorials, and anomalous cases—all of which fall within an occult register while at the same time drifting through infrathin movements away from fascist paranoid subjectivity, sovereign decisions, and the spatialization of the exception in the form of the camp. Indeed, we would go so far as to argue that whereas the camp is the spatialization of the exception from above, the studio is the spatialization of the exception from below (and thus contains within it an antifascist potentiality). Furthermore, McCaffery's criticism of the pataphysics of the camp equates darkness with negativity. The implication is that such darkness must be overcome through a process of throwing light on the subject. Yet as Moten argues, darkness is complex and not merely the result of surveillance and terror. Deleuze's dark precursor is the way the subject perceives the d-rift of the schizo-desire of neurodiverse blackness. Thus, darkness need not be coded as negative. Instead, the valences of darkness experienced in the studio (as a state of exception from below) is precisely the motor driving research-creation. Because McCaffery fails to look into blackness and see nuances of darkness (thus diming his white gaze to accommodate it to low-light vision of black light), then he fails to avoid the fascism he criticizes.

But rather than an ontology to the side (of whiteness), blackness, on our reading, is less a paraontology than a *pataontology* with its own pataphysical alchemy. A black pataphysical alchemy would imply a more radical ontological disturbance of the space-time of whiteness and its algorithmic insistence on capture than even paraontology, which, in the end, still takes ontology *too seriously* as its foil. Furthering Toscano's call for a return to the antifascist theories of racism found in seventies black radical movements,[73] we argue for a similar engagement with black conjuring arts to support an (ae)aesthetics of black, antifascist pataphysical alchemy.

Black Pataphysical Alchemy and Afrofuturist Studioing

Before exiting this gate, we can now return (ouroboros-like) to where we began, to Sun Ra's Afrofuturist film *Space Is the Place*, which we consider one of the ur-texts of black pataphysical alchemy and antifascist studioing practices. In this way, we return to a theme first introduced in the initiation to this book: the a-disciplinary space of the alchemical studio between *Oratorium*, *Auditorium*, and *Laboratorium*. It was not uncommon for medieval alchemists to draw analogies between the manual work of the laboratory, in which certain steps toward the philosopher's stone had to be followed, and the rules of musical composition.[74] Taking up this analogical relation between music and alchemy, we turn to Sun Ra as an esoteric alchemist who pursues transmutation through Afrocentric sonic experimentation. Or perhaps we can argue that Sun Ra is an early manifestation of what Reynaldo Anderson and Charles E. Jones refer to as astro-blackness, which is an Afrofuturist concept emphasizing a post-digital, cybernetic, diasporic black consciousness liberated from colonization and attuned to the potentialities of cosmic intensities through sonic alchemy.[75] Afrofuturism is, in a sense, a direct challenge to what far-right activist and writer Guillaume Faye describes as the *archeofuturism* found in various European fascist movements, which is predicated on visions of a return to an archaic yet futuristic Nordic homeland of Thule or Atlantis.[76] Situating *Space Is the Place* as an alchemical experiment in relation to black, antifascist thought and politics is particularly apt considering that the film was made after Bobby Seale and the Black Panther Party invited Sun Ra to the Oakland in the early seventies. It is our contention that the antifascist political philosophy developed within the Black Panther Party emphasized by Toscano and hinted at by Harney and Moten is transformed into a pataphysical experiment in sonic alchemy by Sun Ra.

Throughout the film, Sun Ra perfectly embodies a cross between Egyptian alchemy and pataphysical paradox, poetry and science, mythology and science fiction, or what he refers to as "myth science,"[77] which enables survival in (and against) various states of oppression through avatars, teleportation, and most importantly sonic experimentation. His music is a ritualizing practice emanating from an (impossible) future that does not recognize strict divisions between knowledge or pleasure, art or science, unleashing new, illegitimate flavors/tastes that multiply through the dissonances that clump together between big band swing, trippy Moog synth, free jazz experimentation, parodies of commercial jingles, and African polyrhythms (just to name a few). Likewise, the film is genre bending: documentary-style concert film, alien science fiction, blaxploitation comedy, and biblical epic all drift in and out of one another. As *Space Is the Place* reveals, the inoperative work of the film creates a sonic time-space machine to transport the audience to another, pataphysical dimension that embraces the antifascist drift of black neurodiversity that has both historical and cosmological dimensions.[78]

At this point, we can return to the opening sequence of *Space Is the Place*, which offers a comparison between on- and off-planet sonic environments. The Earth's soundtrack is, as already noted, a composition of "sounds of guns, anger, frustration." It is pure sonic agitation that is out-of-phase with the cosmos, a stark sonic signature of a cancerous body without organs. Such sound lends itself to hard and cold listening and paranoid (fascist) thinking. Yet, while wandering through a strange forest on a distant, unnamed planet, Sun Ra states, "The music is different here" and "the vibrations are different." The Earth is dominated by the sounds of death, while the haunting, alien vibes of the opening score to *Space Is the Place* offer an alternative, or what Kodwo Eshun describes as "polyrhythmazes."[79] The soundtrack of the Earth is one-dimensional. It has one tone, pitch, and rhythm: the marching orders of destruction. Alternatively, Sun Ra's Arkestra invents polyrhythmazes or labyrinths of sound that play with thresholds between dissonance and harmony, music and noise, aesthetics and (an)aesthetics, always gesturing toward what cannot be heard within what is heard. Polyrhythmazes do not offer resolutions to these tensions they unfold. Instead, the tensions multiply and diversify, becoming mazes in which the ear loses its orientation. Aural disorientation does not lead to pain but rather the "transmolecularization" (as Sun Ra would say) of the ear as a sense organ so that it becomes sensitive to dark frequencies either above or below taken-for-granted sonic bandwidths. There is no resolution to Sun Ra's polyrhythmazes, no final harmony or unity is achieved. Polyrhythmazes offer up a different vibrational set of frequencies that gesture toward unexpected flavors, destabilizing the hard and cold body of the fascist. The film is an attempt to tune the audience into these different vibrations so that something beyond fascism can be heard (something beyond the echo chamber of agitation of anger, suffering, and destruction that fascism endlessly repeats), pole-vaulting the senses into an exceptional state of receptivity that can swerve in their sensorial universe toward an "alternative destiny."

In addition to polyrhythmazes, Sun Ra's musical experiments also embrace an (an)aesthetic of distortion. Robert Worby, in his manifesto for cacophony, writes, "Distortion is accumulation, overdrive and mutation at the limits of a system and these extremes bring about change, be that in sound or in society.

Distortion and feedback in sound cause partials to pile up and collide like molecules in a hot gas. . . ."[80] Distortion is not simply entropy. It captures both senses of drift we have been developing throughout this book. It is a swerve in the trajectory of objects, and these swerves clump together, heap up, accumulate into loose masses that do not form closed systems. This is loose music that, in turn, loosens attachments to certain habituated forms of listening. Sun Ra's music is not a composition but a sci-fi composting of drifting, heterogeneous fragments that spiral around one another to form temporary distortions we call music.

Sun Ra's cosmos is a vast "orchestra" in which each of us is "music" (sun, stars, humans are indeed all music "vibrating"). This formulation indicates that for Sun Ra there are no ontological distinctions between organic and inorganic states. Instead, there are only variants of vibrational drift that temporarily manifest themselves in spatial and temporal clumps—so many accidental forms that range from stars to human bodies. These vibrations that drift and clump emanate from a "void" or from blackness at the heart of the cosmos. This is therefore not an empty nothingness/void, but as in the alchemical tradition, the void of blackness is a primordial, virtual ethernity that opens up the possibility for new ways of sensing, feeling, and thinking. As Moten writes, "Nothing is not an absence. Blackness is more and less than one in nothing."[81] This is a nothing that appears as empty only to those who lack attunement to darkness and its vibrations. If Adorno once insisted that that protofascist soundscape of the American culture industry produced hard and cold "regressive listening,"[82] Sun Ra would agree, and also lobby for musical experimentation in order to generate temporal and spatial anomalies out of the fecund vibrational fullness of the black nothingness that demands new listening patterns that are less rigid and more pliable/loose. Simply put, Sun Ra is attempting to produce schizophrenic listening experiences.

Sun Ra's pataphysical alchemy produces singularities that defy what is scientifically plausible. In one of the most spectacular sequences in the film, we find Sun Ra teleported to a Chicago club in the year 1943. As he plays the piano, the music becomes increasingly unfamiliar and jarring. Indeed, music passes over into noise. The camera turns to the audience in the club. Two military men listen, their faces showing confusion and perhaps discomfort. The dancers on stage lose the ability to choreograph their movements to the music. Glasses begin to shatter, a wind blows through the club, and smoke begins to emerge from the piano. People suddenly fly through the air, defying gravity. The musical sound/noise of Sun Ra's piano thus has the power to suspend and render inoperative physical laws of the universe through sonic distortions. As Nabeel Zuberi points out, the music becomes a science fiction "special effect."[83] A rather banal scene defined by light entertainment, leisurely consumption, and mild misogyny is completely upended by the emergence of a sonic dark precursor and its vibrational excess. The typicality of the scene is replaced by an anomalous particularity that is inexplicable to the characters in the film. The emergency clears out a space-time in which Sun Ra can confront his central antagonist—the Overseer, who is the black embodiment of capitalist greed, conspicuous consumption, sexual exploitation, and even protofascist author-itarianism. As a character, the Overseer is cold and hard toward others and is paranoid of Sun Ra's powers. He is a cosmological embodiment of sover-eign force that has been channeled into a capitalist (if not fascist) form. While Sun Ra's music destabilizes social relations (notice how his piano playing was offensive for *military* men in the club's audience), the Overseer's ominous laughter reinforces existing spatial and temporal hierarchies and indeed thrives upon them. He holds power while Sun Ra redistributes power (or at least cre-ates a space-time in which such power must be renegotiated). The Overseer thinks only in terms of existing patterns (projecting fascist fantasies of abuse, exploitation, and expropriation into an inevitable, fated future) whereas Sun Ra offers up particularities that open alternative apertures in the space-time continuum. While the former offers only stereotypes (the pimp, the gangster, the capitalist), the latter seems to defy stereotypes, refusing categorization of the particulars, which his sonic and visual experiments unleash, multiply, and intensify.

It is important to emphasize the multiple dimensions of inoperativity found throughout *Space Is the Place*. Sun Ra's visitation to Earth is difficult for black youth to open themselves up to receiving (let alone comprehending). In another space-time jump, Sun Ra lands in Oakland, California, the 1970s and enters a black youth center. Music is playing, youth are talking, joking around, and hav-ing a good time, all the while surrounded by images of black musicians, intel-lectuals, and radical political activists. Suddenly Sun Ra appears. The youth question him, asking how they know he is "for real." Surprisingly Sun Ra flips the question on its head and admits that he is unreal. But this exceptional state proves the rule as they are *all* unreal in the sense that black people are invisible within a white dominated (capitalist, fascist) society. One must take Sun Ra for real *precisely because he is unreal*. This pataphysical reversal of logic suggests that paradoxical formulations are a way out of white algorithmic calculation and prediction, which attempt to capture possibilities, all the while dismiss-ing impossibilities as useless and futile. In this context, we can read Sun Ra's suggestion that if black youth will not come with him on an Afrofuturist trip to an alternative space-time dimension, he will have to chain them like white colonialists did with African slaves, as another paradoxical inversion, or rather a replacement of one ship for another (mother)ship. As Marcus Rediker points out in his compelling history of slave ships, this brutal technology of domi-nation and fear was also a space-time machine of resistance. For Rediker, life below deck during the Middle Passage was horrendous but also was the site of an emergent collective that was potentiality revolutionary. He writes, "Here, on the ship, was a new community in formation. . . . Here was the alchemy of chains mutating, under the hard pressure of resistance, into bonds of commu-nity. The mysterious slave ship had become a place of creative resistance for those who now discovered themselves to be 'black folks.'"[84] This emergent life is not merely sacrificial bare life (as Agamben might argue), but is also a life in *alchemical* mutation in which metal bonds become transformed into cultural and political formations of resistance and invention. In the bottom of such ships were pataphysical escape hatches into other dimensions. Rediker points out, "On every ship there were various and important forms of culture: singing and dancing (of the self-chosen, not forced, variety), drumming (the entire ship, being wooden, was one vast percussive instrument), and storytelling."[85] This is not in any way to take away from the horrors of the Middle Passage but rather to emphasize the ways in which black slaves engaged in alchemical and pataphysical sonic experiments in the face of ontological annihilation. Against this backdrop, we can situate Sun Ra's gesture toward the slave ship. He is not planning on enslaving black youth to save them from themselves. Instead, he proposes an alchemical experiment in which chains can mutate into processes of becoming-metallurgical. Stated differently, Sun Ra's pataphysics can take a weapon of torture and fear and *render its function inoperative* so that the chains can become freed up for collective experimentation in what black bodies can do.

The inoperative nature of Sun Ra's intervention is further emphasized through a comparison with the Overseer. In a discussion with black youth in Oakland, the Overseer asks "You mean to tell me a smart little brother like you believes in that shit [Sun Ra]. I mean, here comes this dude from outer space, wearing clothes you never seen before, laying down a rap you never heard and that you can't even relate to. Come on, you're smarter than that!" To which one of the young men responds, "But he hasn't yet betrayed his black brethren to the exploitative, racially and culturally co-opted Caucasian power structure." The Overseer thus presents himself as a "real world" way to uplift black youth through assimilation to capitalism and its codes of success, prosperity, and consumption. From the standpoint of the capitalist, Sun Ra appears to be a true alien offering nothing tangible, only useless/vestigial musical interludes that have no bearing on the lives of black youth struggling in the inner city to survive. Yet as the character Bubba is apt to point out, the Overseer's "solution" is really just a way to co-opt the drift of blackness for personal gain, or perhaps worse, a deeper and more profound inscription of the exploitative Caucasian power structure within black communities. Sun Ra, on the other hand, is a dark precursor of something else. He offers the potential for an event that could shift the very parameters of what counts as "political." And he does so through the play of instruments in a noninstrumental (unrecognizable, weird, queer, disorienting) manner. Or in Sun Ra's formulation, he offers "a multiplicity of other types of destinies" none of which is dictated in advance, none of which is a predetermined telos. This multiplicity speaks beyond the capture of fascist paranoia, which fears such open-ended experiments with becoming and opts instead for a repetition (even in accelerated form) and projection of the same.

Sun Ra's *Outer Space Employment Agency* is equally inoperative and useless. A hard-up, white, NASA employee wanders into Sun Ra's agency only to find that they do not pay wages. Sun Ra tells the befuddled pencil pusher, "We creators never receive for our work." This is a non-alienated form of labor that will not feed back into capitalist cycles of production and reproduction. The nature of the labor itself remains mysterious, a strange, pataphysical/alchemical concoction such as "isotope teleportation," "frequency polarization," "transmolecularization" (a Deleuzian neologism if there ever was one!), and so forth. The white physicist is visually uncomfortable with and flabbergasted by these weird and unfamiliar, pataphysical sciences, quickly leaving. But what he misses out on is the chance to rediscover his potentiality through the study of inoperative works. Sun Ra's *Outer Space Employment Agency* is not *not* an employment agency. It is an employment agency that has lost its function within a capitalist economy. And it is precisely this radical, inoperative state that makes Sun Ra and his employment agency suspicious to the white, paranoid FBI agents who start to tap his phones, follow him, and ultimately kidnap him in an attempt to stop his final concert from happening. Sun Ra is absolutely clear that there is a "death sentence for this planet" and "everyone must die." This message is cosmological but also historically grounded in the particular racial and class hierarchies of neoliberal fascist tendencies, which simultaneously feed on black youth while also tempting them to accept its logic as a "final solution" (in the figure of the Overseer). Against this background, the only real escape is through a radically inoperative, pataphysical departure from the white cybernetic algorithms of capitalist exploitation and expropriation of blackness.

The *Outer Space Employment Agency* not only promotes inoperative works that have no value within capitalist systems and are unalienated/non-wage positions; it also renders inoperative familiar, Earthly definitions of taste and pleasure. A rather stereotypical alcoholic enters the employment agency to ask for a job in order to buy some booze. But he soon discovers that while he might be employed to do "nothing" he will receive "nothing" in return. Unable to earn a wage that would enable his addiction, the character leaves, also confused by Sun Ra's offer for a life beyond cycles of addiction. In another instance, a white woman who is clearly high on some drug asks in a sexually provocative manner if Sun Ra would like to go to space with her. But Sun Ra responds that space is not only "high" but also "low" indirectly indicating that going to space is not simply an excuse (or metaphor) for some kind of trippy escapism. The song playing throughout this sequence of events has the lyrics "If you find Earth Boring/Just the same old same thing/Then sign up with the space age incorporated." Employment with Sun Ra proposes non-knowledge of pataphysical sciences not yet invented and new flavors/affects that are unrecognizable to sense organs trained to respond to capitalist commodities and fascist tastes for power and domination. While capitalism is addictive, it is also "boring" for Sun Ra. He is much more concerned with generating *new* polymorphic and polyrhythmic tastes. Thus, the hard and cold, sedimented sensorial apparatus as a whole is rendered vestigial in order to open it up to unexpected, new relations between knowledge and pleasure.

One more final note on inoperativity. One could read Sun Ra's avatar as a self-deification that put him at the center of his own cosmology. On this reading, Sun Ra would be identifying himself with despotic gods and Egyptian pharaohs, and thus playing into fascist desires for power and domination. Yet, as tobias c. van Veen points out, "Ra does not *identify* with the Pharaohs, but transforms the figure of the Pharaoh just as he transforms himself into an ancient alien deity."[86] In short, Sun Ra does not simply become an ancient ruler/god. He does not embody/channel/conjure their sovereign powers in order to hypnotize the masses to do his (fascist) bidding. Stated differently, Ra *renders the powers of the Egyptian rulers inoperative* through a pataphysical appropriation of their symbols. He makes the symbols *alien* in a figurative sense (meaning they no longer signify in the traditional manner) and a literal sense (as the pharaohs are recast as ancient aliens and their pyramids futuristic technologies). This is not a return to an existing, romanticized African past but rather the blasting of the past out of its familiar context in order to forecast an alien future. At first, this strategy might strike the reader as dangerously fascist, as fascists are famous for unearthing primordial symbols from a mythic past in order to wield their powers for colonialist ends (one can think here of the Nazi swastika and its occult-like force to enchant the German people). Yet in Sun Ra's case, Egyptian iconography is recontextualized, becoming increasingly alien, unmoored from any sacred function (radically heterogenous). Whereas the Nazis attempted to make such symbols operate in their name and in terms of their productive war economy, Sun Ra attempts to render such symbols *inoperative*, queering their signifying powers. Indeed, when Sun Ra and his Arkestra visit the Oakland youth center, none of the youth are impressed with his Egyptian-inspired costumes. Instead, they perceive the exotic outfits as strange and somewhat suspicious (if not amusing). The symbols of Egyptian power and prestige no longer produced reverence for these black youth. Once venerated, the symbols are perceived as profane costumes of outsider weirdos . . . and intentionally so. The symbols that Sun Ra appropriates are not to be worshiped, but rather playfully remixed, reinvented, opened up for new use through alchemical recipes and patarules that conjoin dissimilar elements across time and space. Much like his music, the imagery that Sun Ra surrounds himself with is the performance of an alchemical *discordia concors* that does not trace a clear path back to a mythic past so much as it catalyzes experimentation with getting lost within polyrhyth-mazes. In this way Sun Ra ingeniously reconstructs Egyptian cosmology while at the same time separating it from despotic, auto-theocentrist (fascist) desires.

Likewise, Sun Ra diffuses the signifying powers of capitalist commercial media. When June Tyson sings "If you find Earth boring/just the same old same thing/come on sign up with Outer Spaceways Incorporated," she mimics the typical commercial jingle but with such deadpan humor that its seductive powers are neutralized. While the production and circulation of commodities promise an infinite variety of tastes, pleasures, and indulgences, for Sun Ra and the Arkestra, this is merely a repetition of the "same old same thing" hidden behind the illusion of difference. In both cases, powers to dominate (politically and commercially) are rendered inoperative through sonic and visual tactics of pataphysical syzygy.

This does not excuse the lingering misogyny and patriarchal posturing found in the film. Most troubling is the fact that no women are shown entering Sun

Ra's escape vessel at the end of the film. Yet, if Sun Ra desires to be fascist (as Guattari would argue), this desire is continually troubled by a rigorous practice of producing (an)aesthetic and inoperative works—works that contain a potentiality for rethinking how desire is invested (even beyond Sun Ra's own fascist, patriarchal tendencies). One more piece of evidence can support this antifascist interpretation of Sun Ra: He is decisively *anti-charismatic*, constantly playing while also subverting his role as alien prophet. At times, he appears to almost sleepwalk through the movie, mumbling his lines, hiding behind an affectless mask and paradoxical, pataphysical formulae. Instead of the agitated, aggressive fascist who sways audiences through sheer magnetic willpower and esoteric symbols, Sun Ra allows his message to slowly unfold, drifting (aimlessly, loosely) into the consciousness of his audiences. Indeed, he is the opposite of the fascist leader, sometimes displacing himself in order to highlight the Arkestra. In such cases, the individual power of the ruler-as-God is therefore subverted for a new form of black, sonic solidarity-in-looseness that emerges through the vibrational drift of the music.[87] This, we might say, is Sun Ra's translation of Deleuze and Guattari's art of dosage (so as to avoid rigid territorialization under the sign of the leader or accelerated deterritorialization toward cancerous self-destruction). Furthermore, the cult of the hypermasculine leader of fascism is also rendered inoperative through Sun Ra's queer persona, which resists dominant gender stereotypes, gestures, and affectations. This stands in stark contrast to the Overseer who embodies the typical gangster machismo of male actors in blaxploitation films of the same era. Sun Ra is ambiguous on multiple levels, thus sustaining space and time for queer identities that would otherwise be marginalized (or even exterminated) by the strict gender hierarchies of fascism.

The film *Space Is the Place* is a pataphysical object that continually loops back on itself. Just as polyrhythmazes generate temporal escape hatches into alternative temporalities with their unique vibrational signatures, so too the editing of the film loops back on itself becoming a cinematic ouroboros that ends where it begins and begins where it ends. In the opening sequence, we see a spaceship flying through space accompanied by the chant "It's after the end of the world/Don't you know that yet," followed by images of Sun Ra wandering through an alien forest. The audience watching the film most certainly does not know the Earth has exploded, and that this planet is the destination hinted at during the final escape sequence of the film. Disturbances in the sequential, linear time of the traditional film narrative are further compounded by the way the final concert is interspersed throughout the narrative. On first viewing, the audience is completely unaware that what they are seeing and hearing is a public concert that will happen toward the conclusion of the film. Instead, the jam sessions appear to emerge out of a void of nothingness. The footage is shot against a totally black stage. When interspersed throughout the film, these sequences feel like they are taking place nowhere, or what Sun Ra would call the "Kingdom of Blackness." At the same time, there are certain sequences that appear to place the Arkestra inside the spaceship. This interpretation is given further credence by the fact that the ship is fueled by music. In this sense, the Arkestra is (a) inside the ship, (b) inside the concert hall, and (c) outside in the void *all at once*. There are both temporal displacements (in terms of beginning and end, before and after) and spatial displacements (in terms of inside and outside) throughout the film. Even stranger, it should also be noted that the orchestra is an *ark*estra, indicating that the musical ensemble is its own spaceship (its own ark), and as such cannot be constrained to any space-time coordinates on any existing map. The Arkestra, on this reading, does not need any physical infrastructure (that is, an Earthly stage or otherworldly craft) as it is its own dark infrastructure—it is the *real* spaceship/time machine in the film, embodying what Harney and Moten would call a displaced place of study in a fugitive undercommons. Using our terminology, the musical jam sessions emerging from the Kingdom of Blackness through sonic and imagistic vibrations enact the plasticity of studioing practices as against the rigidity of fascist space-time (and its marching orders).

Sun Ra openly admits that his solutions to fascism on Earth are impossible, pataphysical interventions. He once said, "The impossible attracts me, because everything possible has been done and the world didn't change."[88] Of course, such Afrofuturist experiments in sensorial alteration are inoperative and useless excesses/expenditures (impossible solutions), yet in their very inoperativity, they speak to a form of life beyond fascist "final solutions." They speak of a solution that is improbable. Yet it is precisely because of this improbability that Sun Ra's musical imagination outsteps the capture of white cybernetic probability informing so much of our fascist, neoliberal, accelerationist age.

Algorithmic rules are all about prediction and control of the future. They attempt to minimize risks and therefore minimize the impossible from happening. Indeed, the impossible does not even figure into the logic of algorithmic rules. But patarules are ruled by impossibilities, singularities, particulars, exceptions. These are rules that are unruly rules of thumb and are instruments for invoking dark precursors to rip a hole in the space-time continuum. The impossible "attracts" Sun Ra because he has attuned himself to vibrational intensities above and below common and good sense. Fascism in this sense would not be the emergence of what is impossible, but is part and parcel of the possible. Fascist destruction might feel like the rupture of the impossible within the possible (the horrors of fascist violence are often described as "impossible" or at least "improbable"), but it is instead merely the acceleration of algorithmic logics to an extreme point at which the true costs of capitalism's exploitation and expropriation can no longer be hidden or calculated away as collateral damage. The impossible is a small, inoperative interruption of such possibilities from an outside, from an alien point of interception, or a queer angle of black pataphysical alchemy.

We suggest that Sun Ra's music is the sonic signature of desire itself, or more precisely the drift of black, pataphysical, neurodivergent desire that refuses to submit to fascist (re)territorialization, and thus remains an *impossibility*. There are suggestions in *Space Is the Place* that Sun Ra has figured out a way to convert his "harmonic progressions into energy." Desire is not simply a lack—as in the psychoanalytic tradition—but is also a propulsive intensity or energy that (dis)connects through drift—both in terms of drift as a spreading out and as a clumping together into loose, connective networks, or musical polyrhythmazes. In the final images of the film, Sun Ra's spaceship takes flight as the Earth explodes through what appears to be a series of natural disasters. Dams explode; buildings crumble. The infrastructures built to harness energies of the Earth for capitalist expropriation cannot hold, ending with an explosion of the Earth. Unlike accelerationism, which revels in such destruction as the final unleashing of desire from all territorialization, the ending of *Space Is the Place* offers a warning against nihilistic excesses, conserving the minimal conditions necessary for keeping open a (black) future. But as van Veen points out, the destruction of the planet is ambiguous as the film clearly conveys the message that total dehumanization, commodification, and alienation have *already* happened under slavery,[89] and that Sun Ra's arrival to Earth is *after* the global destruction brought about by racialized, fascist, and capitalist Armageddon and *before* the end of time (the physical destruction of the planet). This is the time of the real state of exception—a vestigial time that remains between the end of chronological time and the end of time in which particularities begin to emerge to take flight, escape pods are released, and a nomadic tribe flees into the desert of space in order to jumpstart a new diasporic drift. Read in this light, the destruction of the Earth should not be considered an impossibility, as its destruction has already unfolded under the accelerationist momentum of deterritorialized fascist violence. And the Arkestra spaceship should not be read as another colonialist voyage of discovery so much as an escape pod of survivors searching for a new beginning.[90] Sun Ra's spaceship thus becomes an exploration of space as the place in which nothing remains except the truly impossible potentialities of drift now returning to the void . . . all that remains is black desire . . . a perfect ouroborotic (meaning erotic and robotic) knot whose singular contraction suddenly expands to the point at which the cosmos itself becomes the dark infrastructure of the studio *par excellence*. Riffing on Ra, we might say: S-pace is indeed *the* place.

Gate Five

Vestigial Research-Creation
and Postdigital Pataphysics

The digital research of the laboratory, simply put, concerns counting. As Florian Cramer argues,[1] the concept of digital comes from the digits of the hand and can thus refer broadly to any research concerned with discernibility, distinction, discreteness, and/or differentiation. The digital therefore has a long genealogy that stretches beyond obvious references to computational systems, including games, textiles and weaving, and other premodern artifacts, objects, and machines.[2] On this broad interpretation, the digital in research is actually redundant as *all* modern, Western, scientific research is digital insofar is it concerns discretization of lived experience into abstracted data points. This has only become more commonplace with the rise of what Daston refers to as the "algorithmic empire,"[3] in which rules-as-algorithms attempt to eliminate all discretion and replace it with calculation. With the advent of data processing, we are witnessing an intensification and extension of digital objects, which are subsumed under calculation, actualized according to algorithms, and networked together by protocols. Yet, at the same time, dark excesses are produced by such algorithmic empires that exceed calculation. For us, *postdigital* is a term that refers not simply to the undeniable implosion of the digital, biological, and cultural spheres but also to the *uncomputable* nature of these implosions.

Through this gate, we argue that the terrain of postdigital research-creation is precisely the terrain of the uncomputable. For us, the modern beginning point for thinking through the uncomputable dimension of the postdigital condition is in the French theoretical interest in expenditure, first theorized by pataphysician Alfred Jarry and then taken up by Georges Bataille. For instance, in his iconic essay "The Notion of Expenditure" written in 1933, Bataille argues that a defining feature of a social structure is unproductive expenditure that cannot be counted or accounted for (cannot be made fully operative or functional). Expenditure is the no-count within every count; it is the dark surplus that haunts algorithmic calculations. Although a turn to Bataille can clarify the unique terrain of postdigital research-creation, it poses a serious problem as well: What are the techniques that would allow us to experiment with, on, and through the postdigital without subjecting it to digital forms of differentiation, computation, and calculability? Methods of digital research are all, in some way, geared toward transforming experience into discrete data points that can be aggregated and disaggregated in order to measure and predict, and to subject postdigital experiences, objects, and relations to these same methods would be to destroy "dark" experiences, objects, and relations that are of interest and concern. As such, we cannot fall back on digital methods in order to understand something about the postdigital condition. Instead, we suggest a new turn toward the pataphysical alchemy of research-creation techniques as a way of accessing the postdigital without thereby destroying it.[4]

In fact, there is a pataphysical demon at the very heart of digital computing that enables algorithmic swerves toward the postdigital. As Andrew Hugill and James Hendler argue, early computer science was inspired by Jarry's interest in "declassifying the classified while simultaneously classifying the declassified"[5] through various games that combined algebraic and geometric formulae with metaphysical and literary play. For instance, in the early 2000s at SpecLab, Johanna Drucker drew inspiration from Jarry and other pataphysicians in order to cocreate a "patacritical demon" as an experimental platform (what we might call a virtual studio) that attempted to intervene into the digital humanities. The patacritical demon allowed Drucker to "think about computing outside problem-solving, engineering sensibility, and to create open-ended, generative capabilities" that would keep the digital from "falling into mere 'instrumental' operations (either administered information management or engineering/statistical calculation)."[6] In this sense, we feel that Drucker's early engagement with pataphysical experiments in digital spaces was an opening toward a postdigital approach to research-creation. Pushing the digital to an even more extreme, postdigital state of inoperativity, Hugill and Hendler point to quantum computing, which accepts the existence of indeterminate phases that are neither zero nor one, thus inserting a dark swerve into the heart of binary logic capable of producing "useless" and "unpredictable" outputs.[7]

Similarly, Matthew Ritchie's pataphysically adjacent work "The Demon in the Diagram" also utilize the figure of the demon to explore the gap that separates and conjoins multiple historical narratives, placing them in a kind of ouroborotic knot. Ritchie plays with the indeterminacy between informational diagrams in the hard sciences with the more imaginative diagrams of fine-art drawings, dramatizing the effects—cognitive and affective—of informational recursion. But what is most interesting in this case is how the demon becomes a role played by the audience itself with the intention to "both gratify and frustrate the urge towards controlled theater that manifests in any visually mediated environment and to collapse the imaginary linguistic severance between forms and signs that might be supposed to separate aesthetic and formal regimes."[8] In both cases, the space that conjoins and separates science from art is where pataphysically dark demons dwell. The demon is precisely that which haunts digital calculations: a figure of what remains of the no-count within algorithmic counting. The demon, or dark precursor, is enlivened through research-creation but only insofar as it appears in and through its disappearance, thus escaping any capture, leaving in its wake glitches, anomalies, and vestigial remnants. In this sense, research-creation within yet against the digital does not merely result in yet another attempt to calculate and know the demon but rather opens a space-time of studioing in which darkness is not abolished so much as inhabited in order to see what kinds of swerves in perception, knowledge, and orientation can be produced.

Throughout the rest of this gate, we will outline two ways in which the pataphysical alchemy of postdigital research-creation differs significantly from the metaphysics of Western, digital research. First, the pataphysical dimension of postdigital research-creation emphasizes non-knowledge through the exploration of exceptional states (dark zones full of demons).[9] Pataphysical research-creation techniques appropriate the form of the algorithmic "rule" but subvert the ends of such rules, releasing rule-bound procedures from their role in calculation and discretization. Instead, rules open themselves up to releasing exceptions, or what Bataille would call expenditure (release the Kraken!). Second, pataphysical research-creation augments the body of the researcher in a strikingly different way from computationally determined research methods, the former producing vestigial organs (unproductive expenditures of the ecstatic body, as Bataille would say) and the latter producing utilitarian, productive organs (for a hierarchically organized, visually oriented body). We then conclude with a brief turn toward David Cronenberg's student film *Crimes of the Future*[10] as a cinematic portrayal of pataphysical research-creation into the noneconomic, nonutilitarian surplus of postdigital reality (in all its ethical and political complexities).

The Postdigital as Expenditure

For Alfred Jarry, Western metaphysics concerns the existence of the general or typical conditions that make reality possible whereas pataphysics concerns the investigation into the "laws of exceptions"[11] that resist any generalization. Science, as the operationalized and instrumentalized embodiment of Western metaphysical assumptions, presupposes as its gold standard generalizability and verifiability yet the pataphysical pursuit of the exception challenges these scientific standards (including the ideal of transmutation in the alchemical sciences from which pataphysics draws so much of its postdigital energy). The exception is precisely that which cannot be generalized (it remains a singular event, eruption, or performance) and fails to be verified (as it cannot be repeated within a controlled setting). Pataphysics thus embraces what science would otherwise reject as nonsense, statistically unimportant, or simply anomalous/ monstrous.

Jarry's emphasis on the exception has been inherited by a number of French thinkers (think of Deleuze's notion of desire, Lacan's notion of the Real, Kristeva's abject, or Lyotard's notion of the sublime). Even if the influence is oblique, one cannot miss Jarry's imprint on figures such as Bataille and his dark concept of expenditure.[12] In "The Notion of Expenditure," Bataille directly draws upon the ethnographic work of Marcel Mauss and his analysis of the structure of the gift—best exemplified in potlatch—in order to critique the Western, classical economic theory of utilitarianism. Whereas utilitarian economics focuses on the calculation of pleasure over pain, Mauss discovered in potlatch an excess of exchange that was not predicated on any notion of personal benefit. Instead, there was an excessive exchange of gifts between groups that produced a series of obligations leading to ever-increasing levels of gift-giving bent on humiliating the other through sheer volume. Instead of amassing wealth, these were displays of epic loss, expenditure, and sacrifice that were absolutely incalculable according to any utilitarian scale of pleasure or pain. For Bataille, Mauss's theory revealed an essential truth about society obscured by the rationalistic approach to exchange found in Western economic sciences: Exchange and consumption are not about production and acquisition so much as noneconomic expenditure. Opposed to any utilitarian theory of the human being defined by needs, Bataille argues that an individual is "not simply a stomach to fill, but an excess of energy to be squandered."[13] The problem with utilitarianist economics is that existence is "not a problem of production and the distribution of products" but rather concerns the question of the "excess energy of workers,"[14] which falls outside of any economic calculus. In other words, expenditure is an expression of pure sacrifice without orientation toward any measurable end (pleasure, profit, utility, and so on).

To summarize, life is not defined in terms of *productive* and *calculable* social activity. Missing from this picture is a series of important events that cannot easily be dismissed as mere exceptions to the utilitarian rule: catastrophes, depressions, orgiastic states, revolutions, all of which are expenditures of such loss that they *break* any utilitarian scale that attempts to measure them in terms of pleasure and pain, gains and losses, positives and negatives. Such events cannot be disaggregated into discrete unites of measure. Indeed, these events and phenomena fall within the realm of the sacred, which, for Bataille, is the exceptional world of inutilious and exhaustive expenditures. In this sense, the sacred stands opposed to not only utilitarian economic theory but also to institutionalized religion, both of which attempt to mitigate the radically disruptive potential of unproductive, impractical, uncomputational surplus.

Digital research concerns itself with what Bataille calls "productive activity" or activity that "serves as a means to the end of production."[15] Postdigital research-creation, on the other hand (or perhaps this turn of phrase is inadequate at this point because hands imply digits and thus an accounting of what is *on* the other hand), concerns "unconditional expenditure" of "unproductive forms" in which loss and sacrifice must be as great as possible. The *unconditional* means that there are no utilitarian laws that place limits on how much will be lost, and *unproductive* means that there is no calculation capable of measuring its outcomes (whatever they might be). The postdigital therefore has no ends orienting it beyond its own massive expenditure (massive in the sense of the mathematical sublime). Bataille includes in his eclectic list of expenditures the purchasing of jewels, cultic sacrifices, games (especially those that end in death), and certain forms of art as unproductive forms of social life. In short, expenditures such as potlatch are already postdigital precisely because they are uncomputable. Within Western utilitarian economic theory and practice (re: digital sciences), expenditure is "rediscover[ed] on the scale of irreducible needs through the efforts of those who push the consequences of current rationalist conceptions as far as they will go,"[16] to the point where expenditure itself is sacrificed (a sacrifice of the sacrifice). Its only remaining expression is, for Bataille, in moments when the bourgeois investment in utilitarian economics breaks apart in a moment of class struggle, also known as revolution. Revolution—the ultimate postdigital event—has been either (a) villainized by bourgeois economists interested in ameliorating working conditions through calculated reforms (increasing profits while decreasing the pain of workers through minimal wage increases) or (b) betrayed by so called scientific Marxism, which attempts to compute it according to irrefutable economic laws. In both cases, the error lies in faith entrusted to *digital* research and theory imposed upon postdigital phenomena. Digital research concerns the secular, whereas postdigital research concerns the sacred (re: the absolutely accursed share of society that has no value, no function, no employment, and no end).[17] Given this distinction, we would nevertheless prefer the term *alchemy* to Bataille's own use of the term *sacred*, as alchemy more directly expresses the *profane and experimental* nature of Bataille's project as a whole (both in terms of transmutations of self and the various inoperative works produced).

At this point we can thus posit the question of the specific *techniques* of postdigital research that can help us touch, contact, conjure up, and study phenomena of (sacred) expenditure. Bataille's own approach to such topics as perversion, festivals, ritual, and sacrifice offer some important reference points. For instance, he focuses his attention on studying precisely those phenomena that digital research either (a) dismisses as uninteresting statistical anomalies or (b) attempts to digitize by imposing its methods of analysis to otherwise unanalyzable and uncomputational exceptions. According to Bataille, bourgeois society—with its utilitarian economics, laboratory sciences, and Catholic institutions—attempts to suppress the excessive (by making it "work" to achieve profit or salvation), yet we might argue that increasingly, this is impossible. Indeed, theorists have argued that we live in a historical moment in which the exception has become the rule, and the law has been suspended, thus making possible the appearance of sacrifice on a mass scale.[18] Various cybernetic attempts to quarantine the exceptional, rationalize it, or control it only demonstrate the growing inadequacies of such digitalizing approaches to contain the increasing levels of expenditure that define political, social, and economic realities. Despite the continual functioning of white cybernetic surveillance systems, acts of terrorism continue to "surprise" us precisely because such events are uncomputable surpluses. Thus, we live in postdigital times in which the explosive nature of the expenditure has become the norm to the point where the division Bataille himself pointed out between the productive and unproductive differences in social phenomena cannot hold. This means that the postdigital is not a specialization so much as a much needed and urgent general study of expenditure on a global scale. Second, Bataille's philosophical writing style does not submit expenditure to overly rational presentation (in the form of logical proofs or carefully plotted, evidentiary claims) but instead attempts to poetically convey an intuitive sense of the uncomputable dimensions of a phenomenon. The key here is that the uncomputable can never be fully present in writing, can never be fully inscribed into a symbolic order, but this does not mean that its darkness cannot be obliquely and fugitively glimpsed through certain techniques of writing. In this sense, the form of the writing is the content. These are starting points, but in the next section we hope to elaborate further.

Rules for Non-Knowledge

Digital research utilizes discretization to produce accurate models for predictive purposes. It attempts to control for variability and minimizes contingencies in order to discover consistent patterns through computational programs, rules, and games. Its space-time is that of the generic, contextless, empty, sanitized laboratory. Postdigital research-creation of expenditures focuses on phenomena that lack coherent models and thus cannot generate consistent patterns. It is situated in the space-time of the rituality of the alchemist's studio in which dark precursors curve, swerve, and multiply exceptional states. The former concerns the accounting of what can be counted whereas the latter concerns the eruption of the non-countable. In this section, we argue that postdigital research-creation draws from pataphysical inheritances in two senses. As previously stated, we see pataphysics as comprised of the following key dimensions: it is a science that pursues non-knowledge of exceptions in order to discover/produce/create impossible solutions. The linguistic experiments of the Oulipo pataphysical group in the 1960s have taken up Jarry's eccentric mix of science, mathematics, and literary experimentation. For Bataille, the way to research expenditure was through ritualistic enactments (more religious than anything else) while for Oulipo, such expenditure was the product of playful appropriations of mathematical and logistical formulizations (exploring the paradigmatic within the algorithmic). In both cases—ritual and mathematical—rules, or sets of constraints, were used to ignite swerves leading from knowledge to non-knowledge.

Ritual, for Bataille's antireligious religion might be thought of as an impractical practice for rendering inoperative the ends guiding bodies, actions, and thought within society. Bataille's secret society named Acéphale—which lasted for less than three years (beginning in 1937 and ending in 1939)—enacted a series of complex, nocturnal rituals meant to evoke exceptional states of Dionysian ecstasy and revery existing in surplus of diurnal, socially normative pressures to conform. For instance, initiates of Acéphale would receive an envelope containing instructions giving the day and time for a rendezvous in the forest of Marly (usually at either the ruins of the fortress of Montjoie or a famous oak tree struck by lightning). The instructions might also include a number of other procedural details meant to separate the individual participant from his/her worldly habits and everyday state of mind. Here we cite a description of an initiation ritual into the secret society Acéphale written by Patrick Waldberg:

As soon as I arrived in Paris I was taken by Bataille up to the balcony of the building where he was living, at 76 bis rue de Rennes. It was dusk. He turned me to face the east, in other words towards the night, and made me take an oath of silence. The initiation I was to undergo was due to take place a few days later. To this end he gave me a timetable and the drawing of a map. On the appointed date, the night of the new moon, I was told to take the train from Saint-Lazare station to Saint-Nom-la-Bretèche. If in the course of the journey I happened to come across any people I knew, I was advised to ignore them, just as, after we got off the train and while we were following the path through the forest, if the same people were also on that path our instructions were to keep our distance and remain silent. The long walk along sunken paths, steeped in the damp smell of the trees, took us in pitch darkness to the foot of an oak that had been struck by lightning, on the edge of an étoile, where soon enough there gathered a dozen still and silent shadows. After a short while someone lit a torch. Bataille, standing at the foot of the tree, took an enamelled dish out of a bag and put a few pieces of sulphur on it, which he then set alight. As the blue flame sputtered, smoke rose up and wafted towards us in suffocating gusts. The person holding the torch came and stood on my right as one of the other officiants walked towards me, face on. He was holding a dagger identical to the one brandished by the headless man in the effigy of Acéphale. Bataille took my left hand . . . the person holding the dagger pressed its tip into my forearm and made a cut several centimetres long, although I did not feel the slightest pain. . . . Someone then tied a handkerchief around the wound, my shift and jacket sleeves were rolled back down again and the torch was put out.[19]

Similar ritual enactments took place every month on the night of the new moon, and were meant to "establish new rules for living."[20] The intervening time was divided into two periods: periods of tension in which members of the community were required to remain silent and had to avoid seeing one another and periods of license, which included every excess. These ritualistic experiments on the self and on community were always to invoke the explosive point of indistinction between life and death (ecstasy) in opposition to the servility imposed by the Christian Church and fascism (both of which attempt to appropriate expenditure for the conservation of their own institutional or dictatorial forms of power).

With World War II looming and with the breakup of Acéphale, Bataille continued ritual practice, this time in the form of meditation. While the details remain obscure, Bataille does describe successive meditative practices, starting with meditation on peace followed by a second meditative concentration on a poem with an insistent rhythm. Next, Bataille would concentrate on visual representations of cosmic annihilation. After several further steps, Bataille would conclude by meditating on grotesque images of extreme torture. In sum, the practice was meant to produce a breach in Bataille's psyche (an existential decapitation, perhaps).[21] In terms of postdigital research-creation, it is important to note how much Bataille struggled to interrupt, suspend, and render inoperative instrumental ways of thinking about the body and life in general. While digital culture turns toward mediation and new ageism to "optimize" one's happiness or to make one more productive at work, Bataille turned toward meditation as a way to explode the constraints of the self and expose the self to an unbearable cosmic expenditure beyond the horizon of utility (economic calculation) or spiritual salvation (religious indoctrination). To do so, he needed rituals that acted as dark infrastructure to support his alchemical experiments. Indeed, we could argue that Bataille's nocturnal rituals transformed the entire forest of Marly into a studio, suspending its common, everyday function as a recreational park into an experimental zone for inviting dark precursors to arrive and fundamentally transform the bodies and perceptual capacities of participants.

While Bataille drifted more toward occult mysticism and religion (certainly religion that would be considered antireligious by Christian standards[22]), pataphysical research into expenditure can also take on more "scientific" or "mathematical" guises. Jarry refers to pataphysics as the science of the exceptional whose meaning and function rest beyond computation (utilitarian or otherwise). At the same time, Jarry proposed mathematical proof of God's existence and used up-to-date scientific research to create blueprints for an absurdist time machine (as discussed earlier). In this sense, Jarry did not turn his back on science or mathematics (as in Bataille's occultist practices) so much as he playfully engaged with them in decisively uncomputational ways. Instead of surrealism—with its rejection of all forms of rules for the free, uninhibited play of the unconscious—pataphysicians such as the Oulipians opted for what might be called *sur-rationalism*,[23] which attempts to produce exceptions out of the systematic use of computational and algorithmic constraints. Key here is the difference between rule and constraint. As Marcel Bénabou clarifies, ". . . people accept the rule, they tolerate technique, but they refuse constraint."[24] Rules are accepted as natural to those who use them (and are used by them!) on a daily basis, but a constraint is rejected as a "superfluous redoubling of the exigencies of technique" therefore becoming "exaggerative and excessive."[25] Instead of rejecting the definition of constraint as the unnecessary and gratuitous redoubling of the rule, Bénabou accepts it without the attending disgust and contempt. Precisely by redoubling the rule, the constraint becomes a technique for researching inutilious and uncomputable expenditures—*patarules* set gestures, ideas, language, and bodies adrift. Like Jarry, Oulipian writers and artists were inspired by ancient geometry, alchemical recipes, and forgotten/obscure literary forms while also inventing new, experimental sets of patarules/absurdist and extravagant constraints that push beyond the very same traditions. In both cases, the constraints of displacement, substitution, addition, subtraction, multiplication, division, deduction, and contraction produce swerve effects on multiple phonetic, syntactical, grammatical, and lexical levels. The dark freedom of excessive expenditure is not born from mere chaos but rather through the (mystical: Bataille or sur-rational: pataphysical) use of constraints, which push rules to their point of maximal absurdity.

Examples of some of the more famous Oulipo, pataphysical constraints include the lipogram (writing that excludes a particular letter), recurrent literature (writing that contains a rule that enables one to pursue the production of the text to infinity or until exhaustion), and the S+7 method (writing that replaces each substantive with the seventh following it in a given dictionary). As these examples demonstrate, for Oulipo there is no distinction between poetry and mathematical formulization. For this reason, pataphysical experimentation is inherently postdigital. It accepts that there are no clear distinctions possible any longer between the arts and the sciences, between creation and code. Oulipo does not deny this but rather maximizes the constraints already at work within language in order to create radical swerves within yet beyond the very same codified communication systems. Even if computers are used to calculate combinatorials for pataphysical experimentation, for Italio Calvino,[26] the aid of the computer is decisively postdigital insofar as it unleashes a strange, unanticipated, sometimes ludicrous set of combinatorials that oftentimes heightens the effects of absurdity, paradox, and useless expenditure.

Whereas digital research can produce productive knowledge that functions within society to make improvements, poetry and the arts could be thought of as postdigital research for producing non-knowledge beyond socially accepted norms and linguistic rules (defining the proper vs. the improper). Knowledge is, according to Bataille, a "stable domain" where "one recognizes oneself, where one recovers oneself."[27] The stability of knowledge means that it can act as a consistent ground, an ontological foundation, for defining, measuring, and assessing phenomena. For Bataille, non-knowledge is unemployable within such a utilitarian framework. With non-knowledge, there is "no guarantee of stability."[28] It reveals that the foundation of knowledge always includes an excess or uncomputable expenditure found at the very limit of the knowable. There are two implications of Bataille's argument. First, non-knowledge is not the opposite of knowledge (antiknowledge). Instead, it is the outside of knowledge *inside* of knowledge. It is the expenditure of knowledge that knowledge does not know about itself (error, cannot compute!). Second, if research remains within the boundaries of what can be counted (and thus counted on as reassuringly stable), then such research enslaves us to what falls within the limits of good and common sense. Bataille thus warns that "at the base of all knowledge there is a servility."[29] There is an ethical injunction at the heart of postdigital research-creation: Rebel against knowledge in the name of experimental freedom!

Embracing non-knowledge means playing with the limits of knowledge by using rituality and sur-rational experiments. These experiments might utilize rules and procedures familiar to digital research, but for Bataille, the game becomes what is "the undefinable, what thought cannot conceive"[30]—an engine or machine for producing what cannot be thought as stable, verifiable, and generalizable (non-knowledge). In this way, the game (of exceptional constraints and patarules, rather than normative rules) is at the edge of knowledge, pointing toward its surplus.

Poetic games in particular have a capacity to invoke a senseless sense of knowledge. Bataille writes, "All communication among men is rich with garbage."[31] While digital research attempts to clean up communication (reducing interference or glitches so as to produce maximally functional, logically consistent systems or precise, replicable results), postdigital research not only accepts the existence of such garbage (useless excess) but actually allows it to exist and even finds joy in its unproductive production. It is scuzzy research, glitchy research that pushes programs, rules, and games to the point where they break down communication to reveal states of non-knowledge that are equal parts exceptional and particular (nonreplicable and irreducible). In this way, programs and games (in both ritualistic and quasi-scientific guises) can be retooled as techniques in what Bataille refers to as "a struggle against the spell in which useful language holds us."[32] They can be deployed to "interrogate the limits of being,"[33] or the waste that is sacrificed by the digital in the name of consistency, security, employability, and productivity. Poetry is trash, waste, but for this very reason it is precious and fragile, needing the special attention of the postdigital researcher.

Non-knowledge, to use pataphysical language, offers up only "imaginary solutions."[34] Imaginary solutions are not perceived as real solutions within digital research because they are not statistically valid (they concern outliers and accidents). Nor do they lend themselves to rational reforms of existing systems. They are also imaginary in the sense that it is unclear what end (if any) they serve, producing instead an emergent form of life beyond measure and beyond known ends. Digital research is guided by specific ends defined in advance by the economics of utility, but postdigital research does not have such firm foundations against which it can define its progress (or regress). How do we evaluate postdigital and pataphysical attempts at escape? How do we measure its success or failure? These questions seem to be rendered inoperative in the face of the postdigital, pataphysical world, which is full of awkward, unprofessional, and maddening risks beyond measure.

The Acéphalic Researcher

Digital research uses various technologies (such as microscopes and telescopes) to improve upon, extend, and intensify the range and scope of the finite capacities of the human body to perceive the world. For Merleau-Ponty, the human body is defined by a certain horizon that sets spatial and temporal limits. Within a given situation, the body strains to achieve what Merleau-Ponty refers to as "maximal grip"[35] or the optimal conditions for perceiving a phenomenon/ achieving an equilibrium between the body and its environment. Digital science enables the body to extend the range of what it can grip maximally. What can be seen and then thought about through the microscope or the telescope is increased exponentially, as is the number of phenomena that can be quantified and calculated. Indeed, a central aspect of digitalization is the discretization of experience. The technologies of digital research fine-tune discretionary capacities, enabling research to count on infinitely small and infinitely vast scales, resulting in undermining objects (reducing them to parts or even particles to be counted) in controlled laboratory spaces.

One can almost think of digital technologies as providing prosthetic organs/ limbs for achieving ever greater abilities to see and calculate the world. Through various prosthetic additions to the body, finer and finer discriminations appear as affordances for counting and calculating. Result: the pixelization of the world. Before the advent of computer screens, scientific grids and algorithmic tables produced through technologically enhanced research transformed phenomena into digital data banks ready for extraction. The prosthetic here is not merely to make a fragmented body whole again, but rather to enrich what a body can do in order to radically alter the firmness of its grip on the world and its phenomena.

In particular, it would seem that Western digital science has first and foremost privileged the head, and in particular the eye, for prosthetic enhancement (think here of Sohn-Rethel's insistence on the separation of the head from the hand so important for the birth of capitalism and its sciences). Although mind and body have been thought of as separate and thus as separable in Western philosophy, sight has also been historically associated with mind, hence the intuitive familiarity of phrases such as "I see what you mean." Making the hierarchy of the senses in Western digital science clear, philosopher Michel Serres writes, "Many philosophers refer to sight; few to hearing; fewer still place their trust in the tactile, or olfactory."[36] Philosophy invests in vision as superior and uniquely connected to mind/thought/cognition precisely because of its discretion: It can view the world from a distance without becoming entangled with the world. Vision produces what Donna Haraway might refer to as the "God Trick"[37] of Western digital science: It has unmediated, transparent access to the world *as it really is*. Bataille undercuts this association between the human, the mind, and the eye. He writes, "Too long has human life served as head and reason for the universe" identifying the hierarchical organization of the body as a sign of "servitude."[38] Instead, the human must escape the head "like a condemned man escaping from prison."[39]

Postdigital research-creation into expenditure must produce a different kind of body. Digital research is predicated on maximal grip through heightened utility and accuracy of vision via prosthetic enhancement. Yet the postdigital is not concerned with discretization so much as with nondiscretionary expenditures. Postdigital research-creation would need to intentionally interrupt, suspend, and render inoperative the body's tendencies toward maximal grip, creating an inoperative rupture in what a body can do, sense, and perceive. What we are after here is not prosthetic enhancement so much as the production of vestigial limbs, or limbs that do not have clear functions, that do not seem to have utility, that themselves are bodily expenditures without clear functionality. It is, as Deleuze and Guattari would argue, a body without organs, or a body on the brink of radical disorganization. This is a body that thrives in the drift between non-knowledge and weird new tastes/flavors beyond measure. The vestigial body is also a queer body. For Sara Ahmed, a vestigial organ is a part that has lost its functionality and yet lingers. This leftover remnant is not, in Ahmed's interpretation, characterized as injurious so much as indifferent: "What is useless neither helps nor hinders an organism"[40] and for this very reason is open to queer appropriations, inventions, and experimentation. Precisely because of its inoperative, excessively indifferent state, the vestigial part is full of destituent potentialities, which, as Ahmed points out, "have more room to roam, to vary, to deviate; to proliferate."[41] In and through the pataphysical alchemy of postdigital research-creation, the body of the researcher undergoes a series of transmutations. Operative organs suddenly wither and wilt (opening themselves up to new uses), while other, vestigial organs sprout. The goal here is to increase variations in what a body can be beyond the capture of digital calculation.

Postdigital research-creation's first tactic is to cut off the head to create an acéphalic body. Instead of a return to a premodern, organicist recomposition of the head and the hand, for Bataille, we must forge another kind of body that is not beholden to the abstract, reified, algorithmically controlling head. Once freed from the tyranny of the head, the body can then experiment with non-existing organs, vestigial organs that are nonfunctional—generating a new kind of thinking that is embodied and embedded, generating sensations that are unproductively productive and productively unproductive. For Bataille, the acéphalic body was the symbol of his secret society of the same name. The decapitated body was a symbol for the death of God, and by extension, the secular God Trick of Western digital research. Instead, he concentrated on the body without eyes or ears as a site for research into expenditure.

Interestingly, in André Masson's visual rendition of Acéphale for the journal *Acéphale 2*, there is no longer a focus on the head but on the stomach, which has growing on it (or in it) a strange, vestigial organ that looks like a labyrinth. The labyrinth for Bataille was a central image of expenditure. Unlike a maze, the labyrinth has no solution, but was rather a perpetual problem, a place where monsters dwell. One does not escape from a labyrinth but rather falls deeper and deeper inside. Describing Acéphale, Bataille writes, "his stomach is the labyrinth in which he himself has become lost, and I along with him, and there I rediscover myself as him, in other words the monster."[42] In other words, the pathways of the space-time of the labyrinth cannot be calculated, an escape route cannot be charted. Instead of solutions, one discovers the monster that one always already is (a nonhuman human form that undermines the supposed dignity of the human "man" and the optics of reason). The vestigial organ of labyrinthian flesh is a metamorphosis that makes it impossible to disentangle "man" from "monster." If the vestigial body of the postdigital researcher studies labyrinthian expenditures that have no escape/no exit, then so too the body takes on (grows) labyrinthian organs that are as mysterious as the phenomena that are to be studied. The outside space of the labyrinth is internalized as the guts of the researcher while the internal organs are externalized and made into a public space. Such confusions confound strict dichotomies, producing a *studioing body* that is its own space-time machine, swerving or curling in on itself while simultaneously exploding outward. The labyrinth is a peculiar type of fleshy, dark infrastructure. On the one hand, it slows down circuits of transmission and reception, on the other hand, such circuits are only made possible through labyrinthian detours (think of the structures of the ear and how they curl, enabling noise to become audible through looping, winding canals). Indeed, another symbol for Bataille of the labyrinth is the serpent devouring itself (ouroboros in Greek magic and, later, alchemy). The labyrinth's loose knot multiplies possibilities for new flavors and new sensations through its coils. As the head is left behind, inoperative, the intestine-as-labyrinthian infrastructure is not only a spatial remapping of the body but also a queer pacing: one that is *urgently* tracing lines of decent into the crypt, yet also *slow* in its wayward, loopy meanderings (and therefore diametrically opposed to any sense of algorithmically straight, optimal, and efficient clock time).

Bataille points the way toward the insight that the researcher affects the research and the research affects the researcher, producing a truly corporeal science. But also, this body, now turned inside out (its guts exposed), not only becomes sensational, but also atmospheric. It is important to once again turn to Masson's drawing to highlight how the headless figure's left foot merges with a billowing cloud drifting above a cavernous landscape. The body of acéphalic man is increasingly indistinguishable from elemental forces of nature, further blurring the very distinctions upon which digital calculation rests. This is a body that is literally adrift on clouds (of ethernity?), leaving the confines of terrestrial gravity, and thus is not bound to physical laws so much as creating pataphysical laws of exception. This is a neuroqueer body that is disaggregated, open to the drift of intensities and atmospheres that palpitate its fleshy surfaces, undermining the seat of rational control (the head).

At the same time, the acéphalic body does not merely privilege the hand. Postdigital, after all, can also be read as post-hand. In this sense, the body that Bataille describes cannot privilege either the head or the hand, but must, in a sense, render both equally inoperative. This is not a romantic synthesis or return to a state in which head and hand are reunited. Such options seem to have long vanished. Instead, the acéphalic body must invent new organs out of what remains of a body that has been ripped apart by capitalism. The labyrinthian, vestigial organs of the postdigital researcher therefore cannot act as telescopes or microscopes (they do not produce small or far seeing), indeed such organs are not scopic at all. Instead, they open up the body to a field of non-knowledges (which are uncomputational exceptions, functionless, unemployable surpluses that are nevertheless pliable for new use).

Non-knowledge, for Bataille, is uniquely connected to the dark arts of conjuring a nonfunctionalized, uncomputational, inutilious form of vestigial life. Thus, we find it a mistake when theorists of research-creation such as Charles Garoian focus on the arts as prosthetic limbs.[43] The underlying assumption of prosthesis is increased utility, maximal grip, and improved functionality. An underlying instrumentalism manifests itself when Garoian writes, " . . . prosthetic interconnections, distortions, displacements, and mutual interactions within the representational space of art make creative and political agency possible within social space."[44] Against abstract (digital) space, the space of prosthetically enhanced, embodied interconnections paves the way for "critical citizenship and possibilities for social democracy."[45] Research-creation is thereby put in the service of political ends. The non-knowledge of prosthetic experimentation with supplemental expenditure is subtly transformed into knowledge for social democracy. But what is unique and vital about certain arts is precisely how they produce disutility, minimal grip, and ineffectiveness. While Garoian's theory of artistic prosthetics emphasizes unforeseen connections and filiations between eccentric parts, the theory, on our view, is *too operative* and thus misses the ruptural qualities of alchemical and pataphysical expenditure, which likewise make the body multiply, mutate, and disperse itself but this time through *impotent* iterations that lack the certainty of ends (such as the reassurances of political agency for social democracy). Such (anti)production is not nihilistic but rather places the body in proximity to the kinds of non-knowledge that produce the queer joy of self-dispensation (in which self and world die and are reborn as an-archic, neuroqueer "monsters").

So, what exactly does the vestigial body of the acéphalic researcher "do" in and through such "undoings"? For Bataille, non-knowledge is uncomputable. It is nondiscretizable and nondenumerable, therefore existing on the dark side of knowledge as its accused expenditure. What happens in and with the darkness of non-knowledge that cannot be verified through digitalization (cannot be counted on, cannot be accounted for, cannot be counted *at all*)? We cannot make sense of it without destroying it (as non-knowledge is fragile and fleeting). For these reasons, the *body* of the postdigital researcher is of absolute importance in the process of playing sur-rational games. "I think," ponders Bataille, "it is impossible to talk about non-knowledge other than in the experience that we make of it."[46] Stated differently, intensified affective flows of the body (and its experience) indicate the emergence of non-knowledge. Bataille observes, " . . . the sudden invasion of the unknown can, depending on the case, have laughter, tears, and not only laughter or tears, but other reactions as well."[47] Laughter and tears—comedy, anguish, terror, and ecstasy—are the affective clues that indicate the presence of non-knowledge. Tears and laughter convulse the body, making it uncontrollable by reason (one cannot talk one's self out of crying or laughing) and vestigial (incapable of continuing to work or function in a productive manner). Heads cannot reason and hands cannot continue their labors—both are, in a sense, cut off. Laughter and tears put the body in an unbound state of becoming. In this state, the researcher and the researched, inside and outside, subject and object are indistinguishable and actively adrift along lines of postdigital flight (always with a certain amount of caution and precision so as to avoid the kinds of self-destructiveness that emerge from fascism's contemporary acceleration and agitation).

In sum, we can formalize pataphysical techniques for alchemical and postdigital research-creation as follows:

> Write patarules/constraints that render inoperative defining habits, ways of thinking/acting/behaving so as to insert a division/gap between participants and expected social, political, economic, educational, aesthetic norms, values, functions:
> Ritualistic, impractical practices or sur-rational rules.
> Cut off the head of the researcher in order to experience various states of expenditure in terms of perceptual, cognitive, erotic, and intra-relational dynamics with self and world:
> What personal and institutional limits are revealed by the expenditure?
> What kinds of vestigial organs and/or surplus affects emerge (states of ecstatic laughter, horror, terror, anguish)? What new uses beyond algorithmic computation and functionality become possible when vestigial organs bloom?
> Record the resulting non-knowledge that emerges through processes of vestigial (excessive, impractical, uncomputational) use:
> Non-knowledge cannot be clearly communicated or digitally calculated, instead it must be poetically conjured, obliquely hinted at, circumspectly rendered (always rigorously, given the nature of the phenomenon).

By following such rules, the researcher generates the space-time of studioing. Such rules are essential components to pataphysical alchemy—providing permissions to take risks in the name of creation while also preventing the acceleration and agitation that leads to present-day fascist whizzing. In terms of the former, it is particularly important to highlight hospitable conditions patarules provide for inviting and sheltering dark precursors that would otherwise be shut out of most laboratory conditions. As cited in the initiation to this book, Melanie Remi Yergeau highlights the damage done to neuroqueer and gender variant children by digital sciences and the extension and intensification of the space-time of the laboratory. To control the perceived deviancies of queer bodies, intensive behavioral interventions rigorously applied in laboratory conditions had to be integrated into domestic spaces. This enabled such bodies to be continually surveyed by family members who were employed as proxy-therapists. The home became what Yergeau refers to as a "prosthetic"[48] limb of the behavioral therapy apparatus: a space-time of exerting pressure on bodies to conform to neurotypical ways of thinking and behaving. In other words, the home-as-prosthetic device integrated the laboratory's digital rule of discretization into the total life-world of the "patient." Recovery, in this sense, means continual monitoring, disciplinary intervention, and surveillance in order to make the body recognizable, useful, and productive for a neurotypical (white) world. Opposed to this, we argue for the invention of patarules to construct fugitive studios that experiment with alternative ratios of non-knowledge and pleasure emerging from the expenditures of neurodrift. The studio is a vestigial materialization of such expenditure, providing the dark infrastructure needed for neurodrift to swerve and clump into a form-of-life that prefers the imperceptible over the recognizable, the useless over the functional, and the unproductive over the productive. In this sense, the studio is hospitable to the kinds of drift that the laboratory monitors, measures, and digitizes in order to "rehabilitate" bodies for specific social, political, and economic functions.

David Cronenberg:
Images of Postdigital, Acéphalic Research

To conclude, we will gesture toward a cinematic artifact that, in some sense, embodies the kind of postdigital, acéphalic research we have outlined above. In no way is this artifact to be thought of as a "model," as a model would transform a vestigial (and therefore functionless) object into a productive teaching tool. Instead, think of our final example as a resolutely *inoperative example*, one that can never be actualized. In this sense, we offer the following as an impossible solution to the question of representing vestigial research. *Crimes of the Future*, one of filmmaker David Cronenberg's earliest works, concerns the strange and tedious adventures of Adrian Tripod, director of a medical institute called the House of Skin. The film presents a dystopian future where an uncanny disease has destroyed most of the world's female population. Named after Antoine Rouge, the "mad dermatologist" who discovered the disease, "Rouge's Malady" causes the patient to secrete a substance named "Rouge's Foam" from the ears, nose, or nipples. The foam is a death marker, as it indicates the late stages of the fatal disease, but it is also an alluring substance that invokes a sensual experience for those who ingest it, which Tripod and other characters take part in. Victims of the disease finally discharge a brown fluid from the mouth preceding death, which like Rouge's Foam, has a tantalizing quality. The House of Skin treats sufferers of Rouge's Malady. Tripod himself was a disciple of Rouge, who went into isolation when he contracted the malady and is presumed dead, and it is suggested that the master's experiments may have spawned the disease. While this may or may not be true, it is certain the disease originated, somehow, from cosmetics.

The film follows Tripod as he navigates a series of institutions and corporations in pursuit of the mystery of Rouge's Malady. These include the House of Skin, Institute of Neo-Venereal Disease, Oceanic Podiatry Group, Metaphysical

Import/Export, and Gynecological Research Foundation, which is more an alchemical, postdigital studio than it is a digital laboratory. The bodily, sexual, and philosophical connotations these names call forth echo similar preoccupations seen throughout the film. It is silent, except for Tripod's narration and an array of bizarre noises that erupt throughout. The audio gurgles, clicks, rattles, drones, hisses, sputters, hoots, croaks, howls, chirps, buzzes, pulses, and squawks for various durations as Tripod tells his tale. His tone is highly mannered—serious, calm, erudite, and contemplative, but with a vocal delivery that suggests something unnerving. Tripod's language has a formal, baroque quality to it, and it tends toward the pseudoscientific. He remains dressed in black during the entire film, clad in a turtleneck and long coat, wearing a pair of glasses with thin, octagonal frames. He is a peculiar, mostly dispassionate figure, wandering a clinical, pandemic-stricken landscape as he transitions from one weird professional appointment to the next.

Modernist spaces—largely vacant, perhaps due to the impact of Rouge's Malady—dominate the scenes. Buildings are imposing figures shaped from glass, brick, and concrete, some in the brutalist style. Prim courtyards with large pavers and manicured grasses comprise most of the outdoor settings. These features, combined with a minimalist approach to cinematography, create what film scholar William Beard calls the movie's "oppressively antiseptic beauty," antiseptic in a clinical sense as well as in terms of impersonality.[49] This antiseptic quality contrasts with the mucky, haptic character of Rouge's Malady, so that the moments of bodily emission feel particularly grotesque.

While the film has a plot of sorts, it often seems intentionally aimless. The roving Tripod moves from place to place without any identifiable motivation, other than a vague compulsion to pursue the legacy of his vanished mentor. When we first encounter him, Tripod states that the House of Skin "is undeniably in decline." Two interns, seemingly the only other employees, have taken over operations, a situation for which Tripod cannot account. They attend to the House's last remaining patient in an odd fashion, chasing him around the courtyard, playing with a stereopticon, and painting his fingernails. The patient eventually succumbs to the disease. When Tripod finds him lifeless in the courtyard with brown fluid spilling from his mouth, he kisses the patient deeply to savor the secretion. As the patient is his last link to the House, Tripod is now existentially adrift. Yearning for his mentor and confounded by the collapse of his vocation, he ventures out, encountering an array of bizarre characters and situations. Importantly, these episodes find Tripod engaged in, or privy to, research and assessments that are, we argue, fundamentally pataphysical.

Before we examine these instances, we must also note the film's tone and style. It is boring, in a radical sense. Long silences and minimal camera movements create a languorous pacing as the film proceeds. With a few exceptions, the color palette remains muted. Tripod's staid vocal delivery, which includes long pauses and slow verbalization, also serves to decelerate the film's perceived speed. Characters rarely talk to one another, and there is no captured audio when they do. Their interactions generally feel listless and hollow. While the plot includes disease, social ruin, eroticism, gender fluidity, strange behaviors, violence, and even pedophilia, one leaves the film feeling as if not much has happened, other than that one's time has been arrested for the hour-long duration. Rather than being indicative of a young director's lack of control over his medium, which Cronenberg's technical and artistic abilities would seem to belie, even at this early stage in his career, these features experiment with the viewer's attention. The film cultivates boredom to draw attention to attention itself, heightening the viewer's attunement to the transactions between perception and thought, ultimately forcing one to confront consciousness in its most basic form. To be self-aware, and at the same time indifferent to that awareness, is to experience a paradox, or, at its most extreme, even a crisis, of being. Paradoxes are immune to computation, and here boredom presents itself as a strategy that disrupts the rules of cinema, edging the film into a pataphysical zone. An early, noteworthy scene suggests Cronenberg's intentions. As Tripod, in a close-up shot, watches his ill-fated patient through a window at the House of Skin, he inexplicably removes his glasses, slowly licks the inside of each lens, and returns them to his face. On the one hand, the act is a conflation of the senses—taste, touch, and vision—preparing the audience for the sensuality of Rouge's Foam, but it also presages how attention and visual perception will be challenged by the film itself, much in the same way silence and erratic noise challenge one's auditory perception. The licking of the lenses is also perhaps a comment on the director's "vision," as expressed through the camera lens. Put another way, Cronenberg, like Bataille, seeks to remove the head of the viewer (and director) to manifest non-knowledge. The audience's boredom mirrors Tripod's own in the face of his mentor's disappearance. He is in some sense coming to terms with a world that no longer interests him, thus his almost aimless transition between employers and his general detachment from others. People and their actions are "opaque" to him, as are seemingly his own. The film cannot be described as entertaining, educational, humorous, moving, or really even provocative, despite the fact that it culminates in the kidnapping of a young girl, whose pubescence will be artificially induced for the purpose of insemination and research, or, more accurately, rape. Here we find the inutilious aspect of Cronenberg's endeavor. The entire film is useless, unproductive expenditure. It does not *do* anything, in the sense of traditional cinema. The sacrifice here is that of the viewer's investment—emotional, intellectual, and temporal—in the productive experience of moviegoing, where some sort of pleasure exchange, through the machinations of entertainment, is expected. What is to be gained by watching such a film? How does one measure its effects? Is it even possible to do so? No, because the film itself is a vestigial organ that defies classification and purpose, demonstrating non-knowledge through the actions of its characters as well as through the film's formal composition, mode of storytelling, means of representation, and position as an unstable object of interpretation.

During his various visits and employments, Tripod encounters situations and engages in practices that are pataphysical in nature. His first destination after leaving the House of Skin is the Institute of Neo-Venereal Disease, where he finds a former colleague receiving treatment for a sexual disease he has contracted from his own patient. It is important to note that, time and again, the ethical boundaries between clinician and patient break down. While at the House of Skin, Tripod crawls next to his patient's bed and reaches over his chest to express Rouge's Foam from his nipple, which he licks from his fingers in the subsequent scene, clearly not typical or appropriate behavior for a health-care provider. Similarly, it is suggested that Tripod's former colleague has slept with his patient, resulting in his condition. More severely, Tripod conspires to violate the young girl toward the film's end, in the name of some obscure "research." While certainly disturbing, these transgressions speak to the extent to which Tripod and his companions have detached from general experience and embraced states of exception. They also underscore the absolute need to apply an ethical framework when engaging in pataphysical undertakings.

We learn that the illness Tripod's colleague has contracted causes him "to create puzzling organs, each one very complex, very perfect, unique, yet seemingly without function." As these vestigial organs are removed, new ones grow to replace them. A nurse intimates that the disease may be "a form of creative cancer." The extracted organs are kept in jars, and the colleague becomes "melancholy" when they are not near him. The colleague's body is caught up in perpetual expenditure, extending itself through the generation of nonproductive parts. This excess bears resemblance to Rouge's Foam in that they both represent the inability to contain an abundance. Fluids overflow through orifices; useless organs germinate, crowd the body, and must be removed. These manifestations are incomputable, resistant to the methods of those who study them. The true cause of Rouge's Malady remains a mystery, as does that of the "creative cancer." Rather than providing any real treatment for these conditions, those administering care to the afflicted seem more mesmerized by their illnesses than driven to attain a cure. They have entered the acéphalic labyrinth in pursuit of the non-knowledge that will deliver them from their state of torpor, or exaggerate it to such an extreme that an ontological rupture occurs. In the film's final moments, as he appears to prepare to rape his young victim, we see her licking foam from her fingers, a signal of her impending death, and Tripod remarks that he "senses the presence of Antoine Rouge." Perhaps Rouge is the beast at the labyrinth's center, the man/monster that both creates and destroys.

Other pataphysical dealings take place in the film, sur-rational practices that comprise Tripod's job duties. At the Oceanic Podiatry Group, he is charged with enacting a ritualistic technique meant to "reverse the psychic relapse, now so common, which can occur under intense genetic pressure." It is unclear what this means exactly. The technique itself requires one partner to lie on the ground as another partner stands above. The vertical administers a series of manipulations to the legs and feet of the horizontal, flexing joints this way and that. Shoes and socks are eventually removed, and the standing party places the underside of the other's foot on his forehead. There is no detectable utility to these absurdist motions, yet *something* is happening. The technique is somewhat intimate, in so far as two bodies physically engage with one another, but it is also an awkward engagement that breaks laws governing normal, social behavior, a ritual that renders community standards inoperative. Such violation does not seem to bother the story's characters, who take to this fetishistic routine with vigor. Rather, the audience's community is problematized by behaviors that would be discounted in "real life." Rules for living are redoubled, and constraints develop as typical modes of interaction are exaggerated. Instead of shaking hands in greeting, two strangers play a sophisticated, technical, and public form of footsie. Improper actions are treated as ordinary, necessary even to Tripod's impenetrable research. The film, the vestigial organ, pulls the audience into the realm of the sacred, where norms are rendered inoperative by the film's constraints. There are the constraints that the characters encounter and cultivate throughout the plotline, but the film itself can be understood as the product of a protocol enacted by the director, consciously or otherwise, which might be summarized in the following way:

> Create a film that does not seek to entertain.
> Allow only a single character to speak.
> Use no other sounds, except those that seem unnatural or out
> of place.
> Emphasize the uninteresting.

These criteria could, depending on the director, result in very different forms of cinema. The protocol essentially forces the audience to confront, remain sensitive to, and ultimately reorder expectations for perception and pleasure when engaging with film.

Crimes of the Future is difficult to sit through even for those used to art house productions. It is easy to discount, which is entirely the point. The title gives an indication that laws will be broken. The characters certainly break the laws of the world in which they exist. There are murders and sexual exploitation and abuse. As he plans to kidnap and rape his young subject, Tripod describes his coconspirators as "heterosexual pedophiles, a group specifically outlawed, though increasingly pervasive nonetheless." While difficult to fully ascertain, the film's world does have an ethical framework, which Tripod and his companions intentionally violate. But are these the only crimes the movie presents? Are there not violations of the laws governing our traditional viewing experiences? The film's constraints seek to behead the audience for the purposes of postdigital, acéphalic research-creation into the phenomena of attention and pleasure exchange, transforming the space of the cinema into an alchemical studio. The resulting vestigial organ, the movie itself, is a pataphysical object that refuses to obey the laws of cinema, instead offering constraints that encourage the play of non-knowledge. There is no other way to gain the experience afforded by *Crimes of the Future* than through the pataphysical means it articulates and embodies. How else could it be done? The Hollywood blockbuster gets metricized in various ways—money earned, seats filled, markets engaged, demographics reached, streams obtained, awards won, etc. Popular websites such as Rotten Tomatoes assign a film's quality a numeric rating. *Crimes of the Future* is a foreign body within such an environment. Investigating its purpose requires a completely different strategy, a different method, one that leads to the labyrinth from which the researcher will not escape.

Yet it is not easy to dismiss the crimes depicted in the film, especially the hints of pedophilia. And this returns us to the question of the politics of research-creation and its connections to dark, occult spaces and practices of alchemical studioing. Like Bataille himself—who, Marq Smith points out, was unable to see the whiteness and colonial politics of his own gaze[50]—so too we feel that Cronenberg did not go far enough into his own darkness. The film was not vestigial enough, reenacting fantasies that keep in place certain fascist desires to abuse the exploited through sexual fetishism (and ultimately transforming the body-without-organs into a cancerous mass). As such, the film as research-creation comes dangerously close to the fixed desire of the fascist paranoic personality (as described in gate four) retreating from a queer desire for radically new flavors and tastes. There is drift here, but the headless and handless (postdigital) body does not take the ultimate risk of awkwardness, which would truly make inoperative the kinds of power hierarchies and divisions that remain at work in Cronenberg's film. Although the film renders film inoperative (it has no plot, no "characters," and is rather boring), this inoperativity produces numbness and indifference (bordering on coldness and hardness) in the audience rather than a collective experiment in the surplus expenditure of laughter or tears. As Bataille's dangerous proximity to fascism illustrates, sometimes there are less-than-precise, infrathin distinctions between fascism and antifascism and experiments in alchemical studioing can result in cold, digital calculations of pleasure/exploitation algorithms rather than discretionary tastes for postdigital and truly vestigial expenditures of the heterogenous.

Vestigial Comments

This book is *sketchy*. It's sketchy in two senses. First, it started out as a series of thumbnail sketches that we hoped would form a completed work on "rational occultism" (Benjamin) or "divinatory science" (Agamben) and the space-time of contemporary research-creation. But at some point, we began to question the ambition of this plan, leaving us with an increasing sense that the gaps in our analysis were overtaking the analysis. In the end, it remains unclear if we have indeed safeguarded the porosity, profanation, contamination, and pliability of studioing through our tentative rules of engagement. Yet if this is the case, then what remains is not merely an "incomplete" document, but rather a testimony to what remains in potentiality to think and experiment with. Or at least this is the thought that brings us happiness as we remain awkwardly in the crypt. . . .

But the book is also sketchy in a second sense. It is fugitive, rather dubious, somewhat suspicious, somewhat suspect. It has been written by two wayward intermediatists who have drawn constellations in the dirt and have cast the bones of cryptids. Only to discover what? The "truth" lurking in the darkness or with "answers" to the questions that guide our divinations? This might be a cryptic work, but there is no ultimate mystery at its core that we have exposed. There is nothing to believe in . . . only the inscriptions of experiments in correlating nonknowledges with unanticipated flavors/tastes. Whenever one travels into the crypt and prefers not to decrypt or encrypt, then one is left with practices that do not fit easily into the categories that compartmentalize science and art, knowledge and pleasure. One is left to awkwardly grope around in a profane space for which there is no salvation to be offered or teleology to point the way out. Instead, one is faced with inoperative works that reside in pataphysical hinterlands that cannot be accounted for through laboratory measurements or religious faith. Alchemical and pataphysical works are sketchy.

Perhaps it is sketchy in a third sense as well. Sketchy can indicate the emergence of form from formlessness but also, at the same time, the disillusion of form back into formlessness. The sketch is somewhat an-archic, wavering, drifting, becoming. The moment it stabilizes, it gives itself way. The sketchy work is an accidental form—a more or less. This book is more or less whatever it is. Stated differently, we have given the book a *ragged* shape through our application of rules of thumb that conjured it up (somehow) out of darkness (or perhaps it conjured itself). It is a lumpy concoction or heaping up of references, narratives, concepts, various theorists, scientists, alchemists, heretics, pataphysicians, artists, and so forth, each of which has (to a certain extent) been ripped from familiar contexts and placed in a pile in which they neighbor each other (somewhat uncomfortably yet cordially). This is a drift that mounds, that heaves with a certain heftiness that, at least to these authors, does not feel very stable. The drift is a *loose* tangle of loops that form knots. It is bundle of threads that, out of necessity, are left dangling over the edge of the book, streaming outward, this-a-way and that-a-way. If we have managed to stabilize them into a manageable size for the moment, they are already pulling away at the fragile, fugitive structure of the book. Thus, there is an urgency to studying the drift before it drifts from itself. But this urgency is, paradoxically, a slow haste that cannot help but find itself dawdling in back allies and eddies that turn in on themselves.

The sketchy book does not (cannot!) abide by its own rules of formation. We are uncertain how rigorously we have stuck to the rules of thumb outlined in the antechamber. At times, the rules have helped guide us (orient us) in the book's construction, while at other times the rules themselves have seemed to morph, perhaps veering us toward dangerous backwaters with a certain abandon. But then again, this seems to be precisely the point of the experiment: to risk awkwardness, unprofessional excesses, and even charges of madness. We take on these risks and embrace them.

As a drifty form, the book cannot conclude with any answers or solutions. The whole thing is an exercise in impossible solutions, a pataphysical rubric cube that enjoys the turning of its various sides too much to care about whether or not it is "solved." It can only end where it began: with an ouroborotic twist back to the potentiality that remains between art and science, pleasure and knowledge, hand and head.

***

We began the book with an initiation. According to Agamben, an initiation does not entirely fit within the schema of ritual vs. play but seems to partake in both. It releases rituality from ritual and it releases playfulness from mere gaming so that these potentialities can become something else, can come in *contact* with each other. Furthermore, initiations are often associated with the sacred as opposed to the secular. They index a secret and hidden world only accessible to a privileged few. Yet for Agamben, the esotericism of initiation "is acceptable, only if we understand that the esoteric is the quotidian and the quotidian is the esoteric."[1] Initiation does not hide secrets but gives them away, reminding us that what is most enchanting is that which is so near, so close, that it remains in the dark. Studioing is an initiation into profane illumination, as Benjamin might say. Thus after all the talk of the elusive philosopher's stone as a magical elixir, pure gold, and immortality serum, the book *Gloria Mundi*, written in 1526, reveals that the stone is "familiar to all men, both young and old; it is found in the country, in the village and in the town, in all things created by God; yet it is despised by all. Rich and poor handle it every day. It is cast into the street by servant maids. Children play with it."[2] Initiations into alchemy reveals nothing more than the quotidian nature of the philosopher's stone, which is handled by all irrespective of age, class, or occupation. It would seem then that the initiation profanes any dichotomy between the sacred and the secular, creating a new flavor for that which is otherwise despised, rejected, ignored, marginalized.

We have initiated the reader into an esoteric adventure through studioing— *our adventure* into a paradoxical space-time that is neither secular nor sacred but rather their (profane) suspension. The book is our work, our inoperative work, that does not reveal a deep mystery in the darkness of sacred things so much as remind us of the potentiality that exists everywhere, even in the most mundane and everyday shadows (the shadow = a darkness that is superficial, that lives on surfaces). The book is our own vestigial limb: not really a prosthetic enhancement so much as an awkward/wayward growth for dimming, darkening, dwelling in the fog of life, suspending the workings (even if incompletely and impermanently) of our stylistics of seeing, sensing, tasting, touching, and thinking. The project has left us in the dark, but we have grown accustomed to stumbling around a bit.

But what then is this "conclusion"? Is it a funeral? Again, for Agamben, the initiation and the funeral are unique in that both are threshold conditions in which the very process of signification (rather than this or that signification) is encountered. Our conclusion is thus not a funeral in the typical sense. Yes, we have entered the crypt, but not to be put to rest so much as to experiment with what remains of the remains of signification when signifiers no longer have their power or potency, no longer serve explicit functions yet nevertheless dwell in murkiness, lingering with destituent potentialities. Here, Agamben offers an interesting warning, "The frozen images of pagan gods and the fearsome figures of the astrological decans and paranatellons, whose larval and larvate survival we can trace without a break across the centuries, like the rest of the innumerable signifiers of the past, shorn of their meanings, appear as oppressive and troubling symbols; these are the precise equivalent of the *larve*, these *are* the ghosts which cultures keep alive, in so far as they exorcize them as threatening phantoms, instead of playing with them."[3] How to play with the occult and esoteric inheritances of the past without "exorcizing" them as threats to enlightenment reason *or* simply reanimating them as symbols of lost, esoteric power, mythic wisdom, or premodern forms of salvation? On the one side, we find the progressive, rational, liberal who attempts to bury such sketchy forms of irrational superstition in the crypt. Good riddance! On the other side, we find the capitalist and the fascist, both of whom attempt to reanimate dead spirits by entering into the crypt and drawing out its potency (the first for economic enchantment and the second for political enchantment). Yet what if we were to travel into this crypt and play differently with the potentialities that remain in alchemical recipes? This would be a clarion call for research-creation through crypto-genealogical study of what remains in the remains. Such research-creation would exist on the threshold between initiations and funerals where potentialities are encountered and dark precursors swirl.

***

Instead of a conclusion, we offer "loose ends." Isn't this final section something of a junk drawer in which all the misfits reside, clumped together, in the dark? Isn't alchemy itself a kind of "junk science" better left in the dustbin of history (or at least the story goes)? Yet in the junk drawer, misfits linger, gathering strength in numbers, waiting to be activated.[4] This is not a curated space-time, but there is some order to it, some basic structure that gives it an amorphous, shape. We toss in another "failed" remnant, and it adds to the overall drift, displacing some long-time residents of the clump, which topple over and now find themselves drifting to the periphery. Something is happening here, in the dark. This is a hospitable place for esoteric practices to flourish and for secrets to lie in wait. Junk drawers: space-time incubators for becoming-imperceptible and metallurgical.

Let's return back to Manning's notion of the anarchive, briefly touched upon in the second gate of this book.[5] Unlike the archive, the anarchive gives itself away to a process that is not its own. Reflecting on her collective practices, Manning writes, "After a few years, in a collective state of confusion, having at best been able to anarchive the process of anarchiving, we succumbed to our own messy archive of images, videos, recordings, writings, and artworks amassed over a decade of experimentation on four continents to ask whether the archive itself might be a departure for what else moves through it."[6] For her, the anarchive is the place where the residue of archiving amasses—that which is nonrepresentable in the archive as an (an)aesthetic supplement. For us, the anarchive is similar to the process of drift that defines the space-time of studioing. Both are clumps that have temporary solidity/form yet ultimately drift away from themselves. Both resist being transformed into a closed system or apparatus in defense of the "Truth." But perhaps there is a difference as well, an infrathin distinction. Our book—our study—is darker, not in the sense that it is more gothic (although it might very well be), but in the sense that it concerns a dimming of the apertures of enlightenment so that darkness can become legible *as* darkness, as the amniotic fluid of the dark precursor. Ethernity is not heavenly but subterrain, dripping from every stalactite in the crypt. The anarchive might be cripped, but not necessarily crypted. This is a special work of esoteric nonknowledge and perverse tastes—of alchemical recipes from books of secrets, of incantations from surrealist rituality, of sur-rational protocols from pataphysical sciences (that have yet to be invented). The crypt of studioing is in the far back of the anarchive—it is the junk drawer of the junk drawer, that obscene, profane inheritance of the occult that critical theorists find too sketchy to bring into the light of day. To riff on a line from Sara Ahmed, perhaps the assemblage of mis-fits in the junk drawer is an act of "queering the [an]archive" that enables us to "find uses for what has been declared useless."[7]

In the darkness of the crypt, we encountered ouroboros, our serpentine, posthuman companion, who has slithered along (or has it rolled along?) our side as we ventured through each gate of this book. Ouroboros is a time traveler, looping past, present, and future together into a knot that has enabled us to spring from medieval alchemy to modern pataphysics to postmodern playfulness. Ouroboros is a technique. Ouroborotics is the fusion of alchemy and postdigital technologies (robotics) and thus lies at the uncanny threshold between discriminatory and algorithmic rules of construction, making sure that neither can become an empire. Ouroboros is an aesthetic. It is a taste for paradoxical turnings rather than forward or backward momentum along linear trajectories. Ouroboros is an ontology of potentiality in which the only discovery at the end is the potentiality of the beginning. The beginning decompletes its own telos, resulting in an inoperative work that folds in on itself. Ouroboros is a becoming imperceptible: as it eats its own tail, it becomes increasingly smaller and smaller to the point where the knot is so tight it implodes once again with a certain atomic force . . . the drift catalyzes drifting. Ouroboros is a politics. Importantly, encountering ouroboros in the studio reminds us that an anti-fascist politics is possible even in fascist times. This is perhaps the greatest risk of studioing with a companion such as ouroboros: finding oneself on the edge of the fascist state of exception without becoming hardened, cold, paranoid and manipulative as a consequence.

Junk drawers are crypts that shelter their secrets just as much as give them away to curious alchemists who are not averse to rummaging around in corners where wild things take root and grow sprouted, vestigial appendages. Junk drawers are tiny studio spaces—the nonfunctional, surplus space within the ordered, functional space of a home, office, laboratory, or book. These vestigial comments are our appendages (gangly hangings and drippings).

***

The laboratory produces monsters. Think of Dr. Frankenstein's laboratory in Mary Shelley's novel about the modern Prometheus. In such spaces, we bear witness to the dialectic of enlightenment in which reason turns into myth and science turns into an occult practice. The laboratory, for all its controlled conditions, cannot control itself from this inversion back into myth. The laboratory of pure visibility and transparency, pure enlightenment rationality concretized and instrumentalized, becomes dark infrastructure for incubating tragic nightmares that speak to the hubris of the enlightened scientist now turned mad. Indeed, the mad scientist's laboratory is perhaps the perfect embodiment of the dialectic of modernist space writ large. According to Anthony Vidler, the modern period attempted to "flood dark space with light" and "open all space to vision and occupation"[8] in the name of scientific transparency and hygienic health. Yet, this obsession with light did not eradicate the dark, which merely returned as a "dark space in bright space."[9] Although Vidler does not utilize the language of dialectics in his more Foucauldian inspired genealogy, there is nevertheless a negative dialectic at work in his analysis. The quest for absolute transparency (Le Corbusier) ultimately produces its opposite: an infrastructure of shadowy darkness.

But we are interested in another sort of dark infrastructure, one whose witch's flight escapes such inversions, liberating darkness from playing the role of a dialectical foil. The alchemical studio embraces the dark as a primary (not secondary) phenomenological experience of the generative potentiality of blackness. This blackness is not a mere negation of a positive term, but is the more primordial state of pataphysical ethernity. Blackness has its *own* fullness, its own witchy potentiality that is not subservient to the light but rather comes first. Dark infrastructure is cosmic.

Now, we can shift to Kafka's interest in this more esoteric notion of dark infrastructure. Certainly, the novel *The Castle* presents the dark infrastructure of bureaucracy which swallows humanity in a vortex of processes and procedures without end and without escape. This is the darkness of enlightened Law revealing itself through a dialectical inversion of the rational back into the irrational. But there is another space-time in Kafka's work: the minor presence of studios and the critters that inhabit these spaces and times. For instance, take the tiny corner occupied by Odradek as one example of a minor studio. Odradek is a clump, an alchemical, "accidental form"[10] that is more or less alive and more or less dead at the same time. Kafka writes, "At first glance it looks like a flat star-shaped spool for thread, and indeed it does seem to have thread wound upon it; to be sure, they are only old, broken-off bits of thread, knotted and tangled together, of the most varied sorts and colors."[11] It might appear to be a "broken-down remnant" yet, Kafka speculates, "this does not seem to be the case; at least there is not sign of it; nowhere is there an unfinished or unbroken surface to suggest anything of the kind; the whole thing looks senseless enough, but in its own way perfectly finished."[12] It is a pile of drifted stuff, cobbled together, animated by an occult spirit. It resides in a corner of a typical family home in the place where dust settles and things are forgotten. Kafka reveals the dark infrastructure of bureaucracies but also the dark, occult-like powers of corners to generate weird creaturely concoctions out of destitute, hobo potentialities that are not so much the dialectical inversion of order, reason, and transparency so much as a disregarded, useless, heterogeneous excess that is of no importance to anyone or anything.[13] Odradek is not Frankenstein's monster. Kafka's creature is not a tragic, cursed figure, haunted by a Law that has abandoned it. Instead, it is a joyful presence that could care less about the Law of the family man, occupying His house for weird, impractical practices that are as enigmatic as his laugh. Whereas Frankenstein's monster emerges out of *stitched* together body parts and thus is tightly knotted to a form (that ultimately deforms), Odradek is a *loose* amalgam that can barely hold itself together. Whereas Frankenstein's monster clings to the question of the human and of the father (certainly an Oedipal monster, concerned with the Law of the family man), Odradek is a swerve toward a posthuman, post-filial problematic of witchy desire. Here, Odradek is the embodiment of the animating energies of the alchemical studio brought to life, not by a tortured scientist but by the interacting of clumping stuff which, given the right conditions, can form singular, exceptional forms without clear origin (Odradek has no history) and without clear destiny (it has no plans for the future)—it is simultaneously incomplete and finished: the perfect ontological description of drifts that drift. Odradek is not the creation of a singular man, but what happens in dark corners when men turn their backs and walk away (leaving stuff to start is mysterious drift). Odradek is ouroborotics in motion: a spindle, a turning, a twisting, a looping thing that whips itself up from the potentialities of blackness into a form-of-life without name or purpose beyond its own vestigial little joys.

***

The ellipsis seems appropriate as the grammar of loose ends. In his "punctology," Agamben argues that ellipsis dots "close (and at the same time leave open)."[14] The ellipsis is at the end yet prefers not to end. It is the ending of the end without actually ending. At the same time, it is not simply a deferral either. It is not, in other words, a deconstructivist suspension of meaning for a future yet-to-come. Instead, it is a mark of the potentiality of what remains at the end that does not end. It is a *loosening of the end as it drifts from its end*. Deferral is replaced by an experience of a dark ethernity that does not gesture toward that which is to come so much as turns around and repotentializes a new beginning out of what appears to be an end. In this sense, we want to add to Agamben's analysis, inserting an occult twist. What happens when the ellipsis *curls* or forms an ouroborotic twist? Instead of a straight line, the dots arch backward. A loop is produced. The tongue of ouroboros tickling its tail. The ellipsis becomes a series of commas rather than periods. , , ,

***

There is an old saying that alchemists salvage the philosopher's stone from the dung heap. Alchemists are to be found in junkyards, rummaging around in discarded shit. These are places where inoperative works clump together, where flotsam accumulates. Opposed to the sanitary, hermetically sealed laboratory, alchemists forge their studios in places that are dirty, contaminated, populated by a teaming underbelly of critters (indeed the alchemists themselves become critters of a certain kind). Yet these are the most fecund spots where destituent, vestigial potentialities/vitalities lie in wait.

Perhaps the quintessential alchemists are none other than children. As Benjamin reminds us, children are "irresistibly drawn by the detritus generated by buildings, gardening, housework, tailoring, or carpentry" bringing together things in a "new, intuitive relationship."[15] Like the alchemist, the child uses discretionary rules (always thickly situated in a context, always open to change, always in a state of becoming) to cobble together things to make new uses. They see in the discarded, not remnants of what has been lost but rather the shimmer of what never was but for this reason is still available for use.

Children remind us: Alchemy is research-creation from within the ditch of being, in the muck of blackness.

***

The space-time of research-creation has been a key concern of this book. On the one hand, we have problematized the division between studio creation and laboratory research. This distinction merely reiterates the capitalist division between head and hand, science and art, knowledge and pleasure. We have attempted to suspend and render inoperative these dichotomies by returning to a point of contact between research and creation as imagined through a crypto-genealogy of the alchemist studio (and its weird descendants, including bastard cousins such as pataphysical physicians). On the other hand, we have attempted to institute a division between creation and creativity so as to retain the possibility of creation through encounters with dark precursors (so many impersonal, alien forces from the outside) as opposed to the cultivation of subjective and personal skills/talents/capacities for creative industries, manifesting personal goals/aspirations, and so on.

Drift sometimes discovers points of contact between unlikely bedfellows, and at other times, it creates distinctions where they might otherwise not be perceived.

But perhaps research-creation itself is too strong, too operative. As Agamben argues,[16] alchemy ought to produce *inoperative works*, remnants in which the alchemist is left to contemplate vestigial (destituent) potentials that define a form-of-life that does not abide by the forms of life within capitalist society. These remnants are not creations so much as the muck, slime, gunk of experimentation. Through these "failures," an inoperative work *works* on the order of things, the partitioning of the senses, and on the orientation of the subject. For this reason, perhaps what has actually been at stake in this book is not the space and time of research-creation so much the space and time of research-(de)creation. This would not amount to an infinite deconstruction of presence so much as a tribute to the presence of vestigial potentials that live . . . in the dark . . . in the corner . . . gathering dust. Research-(de)creation would be a form of research (a loop, a knot that returns to its search over and over again in an act of rituality) that reveals its secret to be nothing more than a vestigial potentiality that does not work or fit into the world and for that reason is truly an exceptional work, which is functionless to either capitalist or fascist instrumentalization. Creation of works that de-create themselves.

***

Remedios Varo's painting "Alchemy or the Useless Science" (1958) is, in many ways, a distillation of many of the key themes we have addressed throughout this book (figure 2).[17] Indeed, one could learn a lot more from simply contemplating Varo's painting than from reading this clunky text. But maybe that is being a little too harsh. Instead, this book and this image should be clumped together. For Varo, alchemy is a "useless science." It is inoperative, pataphysically impossible, vestigial. Western science is oriented toward solving problems, improving health, maximizing efficiencies, and measuring outcomes based on the empire of algorithms—all of which cannot abide by useless expenditures such as alchemy. The contemporary interest in new age magick, which appropriates the occult for personal wish fulfillment, likewise downplays the uselessness of alchemy in order to sell a prepackaged notion of spiritual fulfillment, health, and prosperity (not unlike early modern alchemical charlatans peddling recipes for gold). Nor does alchemy align with the popular Jungian interpretation of alchemy as psychoanalytic treatment in which stages of alchemical transmutation signify levels of psychological awareness and growth. And finally, the useless nature of alchemy does not allow it to be easily adopted by fascist accelerationists who desire to speed up certain social, political, and economic catastrophes through esoteric interference. What these all share in common is an interest in instrumentalizing alchemy, making it function as a means oriented toward a specific end. Varo turns her back on such appropriations, emphasizing the uselessness of alchemy.

Figure 2. Alchemy or the Useless Sciences. Note: this is not the full painting. *Source*: © 2025 Remedios Varo, Artists Rights Society (ARS), New York / VEGAP, Madrid. Used with permission.

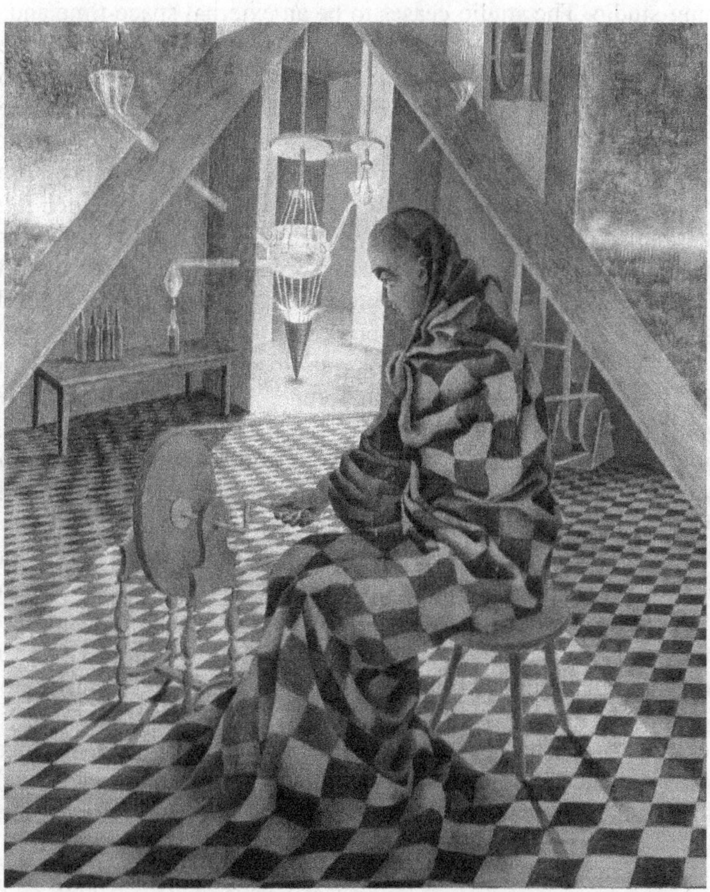

At the same time, Varo's emphasis on the uselessness of alchemy does not mean that it should simply be forgotten as a dead end. The alchemist in her painting is doing something. There is a purposiveness—a slow haste—to the enigmatic actions depicted even if a definitive purpose cannot be pinpointed. Alchemy is an experiment in generating the space-time of exceptional states in which weird loops and loose knots can form out of inoperative gestures, practices, concepts, and materials. Something new, undestined, and destituent can emerge if one can sensitize one's sensorial apparatus to what swerves in the dark. In Varo's image, the alchemist continues a useless experiment *because* of its uselessness (not in spite of it).

The importance of thinking alchemical space-time is emphasized in the painting through the intimacy of the alchemist and the studio. The androgenous figure—who recalls both the alchemical and surrealist interest in the chemical wedding of male and female—in the foreground literally merges with the background, with the space-time of the studio. The patterned checkerboard floor, which echoes the checkerboard floors often rendered in depictions of alchemist studios, becomes fabric draped over the alchemist's seated body and ultimately seems to mix with the very flesh of the alchemist. In this way, the alchemist is a becoming-studio. The studio ceases to be an external space-time and instead wraps around the body of the alchemist. Furthermore, the figure is seated in an ambiguous space that is neither inside nor fully outside. While the studio enclosure opens itself to the outside, at the same time, the experiment of the alchemist is framed by two flat beams that slope downward from the central tower of the studio, both of which make sure that the becoming-studio has a protective circle around itself. The studio is apart from the landscape of which it is a part of. Thus, male and female, body and space, inside and outside dichotomies are overturned in this queer studio that seems to repudiate dualisms. The body folds into the background, becoming-studio, yet the head and hands are still distinct. But these are not unchanged by the process. Because the work performed is useless, the hand turning the enigmatic crank become equally postdigital, vestigial. Although far from acéphalic, the figure's head no longer signifies individuality. It is masklike. Its profile has an indeterminate quality that withdraws from signification. Or perhaps we can even venture that the "head" here is not the human head at all but rather the strange architectural space surrounding the human head and ultimately subsuming it within an almost halo-like, yellow glow. On this reading, the human head would become a residual leftover of the human, replaced by an external, impersonal machine. Thus, head and hand are not so much romantically reunited as they are transformed into vestigial appendages of a new, posthuman apparatus that does not abide by extant divisions but rather invents new, impossible configurations of bodies, landscapes, and atmospheres.

At the center of the image is a complex alchemical architecture/device that is composed of various flasks, tubes, and funnels that protrude from the ceiling of the building and emit an ether. The setup seems to be a fusion of art/handicraft production and scientific experiment yet operating with a logic that is beyond either. This is just as much a surrealist aesthetic as it is a pataphysically sur-rationalist experiment. A fire within a precariously balanced cone provides the heat for the distillation process. Although the fire gives off heat and illuminates the innermost chamber of the studio, the alchemist sits outside this light, in the darkness, focused on turning a wheel that is attached to a gearbox set inside the right wall of the building. The alchemist remains in shadows as if absorbing the light or returning the light to a more basic darkness. There appear to be two substances produced by the churning of the gears within the studio. The first is a green, distilled liquid that drips into bottles. One is tempted to interpret this greenish elixir as the philosopher's stone, promising immortality. But this reading would instrumentalize the work of alchemy, giving it a useful application. Varo is, as the title indicates, interested in a useless science that only offers impossible solutions. As such, what appears to be a useful product of the alchemist's experiment turns into yet another useless byproduct. Also notice that the alchemist produces another, more immaterial substance: a loose, gaseous ether that seems to merge with the turbulent atmosphere in the background of the painting. Perhaps this is a depiction of what pataphysician Alfred Jarry would describe as ethernity? If this is the case, then the experiment and its dark infrastructure form an ouroborotic loop: The work produces the very ethernity that is its precondition, effectively returning again and again to its origins in order to contemplate what remains in potential. The potentiality of ethernity (which cannot be bottled), simply wafts away, and is thus indistinguishable from mere, useless exhaust. Through the unusual technique of decalcomania, Varo creates a textured surface in which foreground ether and background fog press into one another, forming a strangely biomorphic pattern in which various vitalistic becomings seem to simultaneously emerge and withdraw. This is a virtual zone of indistinct figurations, or intensities, that seem to vibrate quietly in the background darkness of the image, never fully giving themselves away. But then again, isn't this the point? Alchemy as a useless science can only produce vestigial by-products/fumes/emissions, but for this very reason, it holds open a space-time that prefers not to submit to the law of useful productivity that dominates both the criticisms of and the attempts to rehabilitate this occult practice.

One note of caution: This studio is apart from the landscape of which it is a part of, thus it is a circle that is as enclosed as it is open to visitations. Yet only one figure is seen in the studio. This emphasis on the individual seems to fall into the trap of imagining the alchemist or artistic genius alone, in isolation. We would prefer an image of multitudes, drifting into the studio space in order to create accidental conglomerates. But perhaps Varo is not romanticizing such isolation so much as problematizing it. If this is the case, then the image is an open invitation to *walk inside its frame* not unlike the unique spell work we described in Saar's multimedia piece titled "The View from the Sorcerer's Window." Varo's painting would therefore be an *initiation* into another space-time. . . . Here, at the very end, during this book's final moments (its funeral, if not burial, in the back of the junk drawer), we begin a new initiation. We feel the checkerboard seeping out of the frame, onto the floor of our studio space, onto our shoes, up our pant legs. . . .

***

What kind of space is the studio? Have we not answered this question yet? Perhaps we can call upon Michel Serres for a moment to make a final claim: that the studio is a "softening box."[18] For Serres, the world of sensation is a world of nested boxes, or more specifically, *black* boxes. As Serres describes, "Take a black box. To its left, or before it, there is the world. To its right, or after it, travelling along certain circuits, there is what we call information. The energy of things goes in: disturbances of the air, shocks and vibrations, heat, alcohol, or ether salts, photons. . . . Information comes out, and even meaning. We do not always know where this box is located, nor how it alters what flows through it . . . it remains closed to us."[19] Another way of putting this is that the "hard" or "large-scale energies" that could "tear your body to pieces" enter the box (so many swirling, excessive, dark precursors) and undergo a fundamental change of state to become "soft" or "smaller-scale energies"[20] that enliven rather than annihilate bodies. Within the box, disturbances enter in and *flow* through it, somehow changed. Notice that such boxes do not prohibit flow but rather facilitate it. And through this flow (or what we are calling drift), a transmutation of something into something else occurs as if by accident or by magic. The flow is a softening process, meaning that it shifts energies, intensities, vibrations to frequencies that are not dangerous for our humble bodies. The black box, as Deleuze and Guattari might say, is a technology for regulating dosages of intensities that could otherwise result in whizzing acceleration. In the box, there is a labyrinth that takes hard energies and diverts them through loops and twists so that nuances and subtleties can diversify and drift out the other side. The darker this transmutation is, the more it resembles the hermetic science of alchemy (a dark art).

The blackness of these boxes should not be underestimated. For Moten, such blackness ought to be read paraontologically as the generative potentiality of blackness. The blackness of the boxes softens cosmic blackness into shades of opacity (darknesses) that resonate with bodies that are raised to be dependent on the light (on enlightenment) and thus are not sensitive to the strange flavors of mist, shadow, and fog. Black boxes are also blacklisted. They are fugitive space-time machines that exist in and produce undercommons composed of studious bodies that do not belong in the white cube of the artist's studio or the scientist's laboratory. Black boxes are dark infrastructure for alchemical experimentation with what must be continually foreclosed upon (but also produced!) by the division between art and science, pleasure and knowledge, head and hand. Thus the "black" in "black boxes" extends far beyond Serres's initial theory. Indeed, we can only understand this blackness in relation to the alchemy of transmutation.

In sum, the studio is a loosening box (Benjamin), a softening box (Serres), a shit box (Bataille), a black box (Moten), a risky/Pandora's box (Dufourmantelle), a neurodiverse box (Manning), an inoperative box (Agamben): a box that does not close in drift, so much as facilitates its reception and transmission.

***

The clock has struck midnight; the gates are closing! So, wrap it up already! But what is to be "wrapped"? How can one wrap up drift when it leaks and seeps through the containers that try to limit it. The box of this book might be too loose or too soft for its own good. At the same time, we refuse to "leave it hanging" as this gesture feels like a cop-out. So, there is only one possibility open for us: to end with an invitation to begin again, to venture a new experiment in producing useless expenditures.

Feel free to use this book (and by use we mean playful research-creation that predates academic functionality or utility):

> If anyone cites this book (Great Oden's beard only knows why someone would do such a foolish thing!), cite it obliquely, with a wink, or a subtle gesture that does not abide by American Psychological Association (APA) guidelines. Or, perhaps only cite the book according to the A'PA (the American Pataphysical Association's nonexistent handbook), which is only an infrathin derivative of the former that nevertheless makes all the difference.
>
> Risk illicit reading practices (drinking games involving absinth are definitely encouraged) that take irreverence seriously so as not to make the book into what it could never have been: a magical text for salvation, truth, divination, prophecy, rather than a sketchy text of a witch's flight.
>
> Proliferate pataphysical centers and alchemical studios in your home institutions (schools, universities, mortuaries, pawn shops, garages, etc.) that study analogical singularities, laws of exception, and impossible solutions. These should not be officially sanctioned, and should actively occupy hallways, basements, storage rooms, windowsills, and empty corners as an invisible, dark infrastructure for studioing that can be taken up by anyone in order to render things, ideas, and gestures inoperative. Risk building out illicit infrastructures that serve no purpose for the institutions they vestigially occupy.
>
> Create a syzygetic contact between concepts in this book and seemingly unrelated sets of concepts (maybe junk concepts

from various academic crypts). The point here is not to hybridize but rather to imagine what exists in the gap that might separate and conjoin singularities. Don't do this in a flippant way, but take care to think and feel in the dark so that the contact points give rise to new questions and new uses. Risk transmutation through rules that drift and drift that loops and knots into new rules for a form-of-life that is more or less an *accidental* form not destined for this world so much as a coming world (a pataphysical world).

Test out forms of antiquated writing (such as alchemical recipes, books of secrets, pataphysical protocols, and so forth) not for the purposes of achieving personal, spiritual growth or professional accolades so much as to see what effects they might have on producing new flavors and new forms of non-knowledge. Risk unprofessional or mad discourses on the fringe of common and good sense in order to produce communities that do not yet have a name (inoperative communities of studiers). Take whatever you have underlined or noted in this book as "important" or "interesting" and redact it all using a dark marker. Next, take whatever "unimportant" or "uninteresting" bits and bobs that remain and clump them together, producing a vestigial or useless text. What kinds of drift does this cause in your desire? Does a new flavor emerge from such a tasting? Appropriate any given algorithmically accumulated set of metadata and transmute it into patadata, or data that serves no functional purpose. Alchemy produces gold *as not* gold (or inoperative gold that has no value as a commodity, offering only the "false" appearance of gold); likewise, patadata is data as not data. This transmutation does not erase the existence of data but merely renders it into an inoperative work that, subsequently, is open for new, queer use. It makes data illegible to the statistical sciences by articulating data through a dark tongue that speaks only in riddles, paradoxes, and studious secrets. Risk profaning the sacredness of digital research in the name of research-creation.

Find the false citations in the bibliography (hint, hint) and write articles with these titles, publishing them retroactively in yet-to-be founded publications, thus inverting the typical linear and progressive time underlying citational logics. Risk temporal paradox.

These are weak interventions, bordering on totally disappearing into irreverent (if not pathetic) trifles. But perhaps this is the problem: we all feel that the stakes are so high that any experiment is stifled before it can take place. How can we practice loosening when we are so tightly bound (knotted) to the immediacy of urgent clarion calls to action? Against all odds, let us posit the following, counter intuitive, pataphysically deviant declaration:

Become alchemical, for the stakes *could not be lower*! Now more than ever we are in need of inoperative works that can be launched in an impotent attack against the titanic force of algorithmic metadata and white cybernetic surveillance.

Give yourself license to *indulge* in useless drift . . . forming clumps of mis-fitted alchemists and pataphysicians whose research-(de)creations in an expanding circle of posthuman studios offer another space-time continuum for antifascist and anticapitalist tastes for shadow flavors and cripped/crypted non-knowledges.

Notes

Initiation

1. Stengers, *Invention of Modern Science*, 84.
2. Daston and Galison, "The Image of Objectivity."
3. Daston and Galison, "The Image of Objectivity," 83.
4. Daston and Galison, "The Image of Objectivity," 99.
5. Agamben, *Stanzas* and *Taste*.
6. Marx, "Inaugural Address," 11.
7. Sohn-Rethel, *Intellectual and Manual Labour*.
8. Sohn-Rethel, *Intellectual and Manual Labour*, 64.
9. For a more detailed analysis of this narrative, see O'Doherty, *Studio and Cube*.
10. Latour, *We Have Never Been Modern*, 35.
11. Latour, *We Have Never Been Modern*, 37.
12. Bishop, "Information Overload," n/a.
13. Loveless, *How to Make Art*, 56.
14. Loveless, *How to Make Art*, 100.
15. Indeed, the "fall of the studio" was first proposed by Robert Smithson in 1968. See Caroline Jones, *Machine in the Studio*, and Caitlin Jones, "The Function of the Studio." Such perspectives condemn the notion of the studio as an exclusive space of the male genius.
16. O'Doherty, *Studio and Cube*.
17. Fuller and Weizman, *Investigative Aesthetics*, 211, 215.
18. Rogers, *Art, Science, and the Politics of Knowledge*, 150.
19. Manning, *For a Pragmatics of the Useless*, 115.
20. Manning, *For a Pragmatics of the Useless*, 329.
21. Manning, *For a Pragmatics of the Useless*, 68.
22. Manning, *Minor Gesture*.
23. Loveless and Manning, "Research-Creation as Interdisciplinary Praxis."
24. We will refer to these practices collectively as research-creation experiments even though this terminology would not be acceptable to all those working at the contact point between art and science.
25. Rousell, "Inhuman Forms of Life," 888.
26. Agamben, *The Use of Bodies*, 237.
27. Latour, *We Have Never Been Modern*, 59.
28. Faivre, *L'esotérisme*, 13–21.
29. See Röhr for more details on this history.
30. Principe, *Alchemy*, 21.
31. Cited in Hartmann, *Life and Doctrine of Paracelsus*, 163.
32. Eamon, *Science and the Secrets of Nature*.

33. Maier, "New Chymical Emblems Relating to the Secrets of Nature," *Atalanta fugiens*, https://furnaceandfugue.org/atalanta-fugiens/frontispiece.html.
34. For a detailed interpretation of the multimodal nature of Maier's text, see Nummedal "Sound and Vision: The Alchemical Epistemology of Michael Maier's *Atalanta fugiens*."
35. Roob, *Alchemy and Mysticism*, 11.
36. Cited in Roob, *Alchemy and Mysticism*.
37. Nummedal, "Sound and Vision."
38. Zosimos, *Of Virtue, Lessons 1–3*, 51.
39. Bacon, *Radix Mundi*, 115.
40. Bacon, *Radix Mundi*, 17.
41. Magnus, *Libellus de Alchimia*, 101, 109.
42. The role of gender in alchemy is more complicated than this simple observation. See Warlick, "The Alchemical Androgyne," for more details.
43. See Morton, *Dark Ecology*.
44. Zosimos, *Of Virtue, Lessons 1–3*, 52.
45. Cited in Forshaw, "*Oratorium—Auditorium—Laboratorium*," 171.
46. Forshaw, "*Oratorium—Auditorium—Laboratorium*," 171.
47. See Kieffer, "The Laboratories of Art and Alchemy at the Uffizi Gallery in Renaissance Florence," 105–27.
48. Two early examples of this can be found in the artist studio spaces of Tintoretto and Michelangelo (Cole and Pardo, "Origins of the Studio," 1).
49. Of course there are many other artists in the modern and postmodern periods who also conceived of their studios in alchemical terms, including Marcel Duchamp (see Moffitt, "Cryptography and Alchemy in Work of Marcel Duchamp and Walter Arensberg"), the surrealist Max Ernst (see Warlick's *Max Ernst and Alchemy*), the enigmatic puppeteer Robert Anton (see Policht, "Divas, Junkies and Alchemical Experiments"), and the queer filmmaker Derek Jarman (see Noble, "The Wedding of Light and Matter") just to name a few. In all cases, the studio was a space-time for radical experiments in the transmutation of self and objects through alchemical experimentations and ritual performances.
50. See the vestigial comments at the end of this book for an analysis of Varo's relationship to alchemy.
51. Janet Kaplan, *Unexpected Journeys*, 96.
52. Kaplan, *Unexpected Journeys*, 74–75.
53. Carrington, *Down Below*.
54. Susan L. Aberth, *Leonora Carrington*, 50.
55. Algazi, "At the Study," 36.
56. See Roob, *Alchemy and Mysticism*, 322–23, for examples.
57. Cited in Eliade, *The Forge and the Crucible*, 163–64.
58. Thornton, *The Scholar in His Study*, 120.
59. Deleuze and Parnet, *Dialogues II*, 9.
60. Mercer, *Alchemy*, 193.

61. Newman and Principe, *Alchemy: Tried in the Fire.*
62. Wood, "Indoor-Outdoor: The Studio Around 1500."
63. Cited in Hall, *The Artist's Studio*, 68.
64. Comenius, *Orbis Sensualium Pictus*, 189.
65. Trismegistus, "The Emerald Tablet of Hermes Trismegistus."
66. Morris, *The Matter Factory*, 30–33.
67. Ahmed, *Queer Phenomenology.*
68. Stengers, *La Vierge et le neutrino.*
69. Sharp, *English Heretic Collection*, 20.
70. Bataille, *Encyclopedia Acephalica*, 35.

Antechamber

1.   Chaucer, "The Canon's Yeoman's Tale."
2.   For an overview of this critical literature (as well as surprising eighteenth-century endorsements of certain alchemical ideas), see Allen G. Debus, "Alchemy in an Age of Reason: The Chemical Philosophers of the Early Eighteenth-Century France," in *Hermeticism and the Renaissance: Intellectual History and the Occult in Early Modern Europe.*
3.   For an overview of the history of Western esotericism see Wouter J. Hanegraaff's *Esotericism and the Academy: Rejected Knowledge in Western Culture.* Although our present purposes differ distinctly from Hanegraaff's, the story he articulates provides evidence of how the head (academic sciences) became separated from the hand (the corporeal science of alchemy) leading to a divided world.
4.   Marx and Engels, *The Communist Manifesto*, 226.
5.   Marx, *Capital: Volume One*, 229.
6.   Adorno, *Minima Moralia*, 238.
7.   Adorno, *Minima Moralia*, 239.
8.   Adorno, *Minima Moralia*, 239.
9.   Metzger, "Operation Mindwarp," 12.
10. Morrison, "Pop Magic!," 28–29.
11. It is useful to employ Christopher Partridge's distinction between "occulture" (which is the non-oppositional, everyday occult references and ideas found circulating in the culture industry) and "occult culture" (which is much more secretive, oppositional, and countercultural) for distinguishing between the two (parallel and at times intersecting) trends discussed here. See Partridge, "Occulture Is Ordinary."
12. Jacob Christiansen Senholt, "Radical Politics and Political Esotericism."
13. See Olivia Goldhill, "The Neo-Fascist Philosophy That Underpins Both the Alt-Right and Silicon Valley Technophiles," *Quartz* <https://qz.com/1007144/the-neo-fascist-philosophy-that-underpins-both-the-alt-right-and-silicon-valley-technophiles> and Christopher Haworth, "Post-Punk, Industrial Culture Zines, and the Information Dark Age."

14. Evans, "Accelerationism, Amphetamine Philosophy, and the Death Trip."
15. See Carroll's *Speculative Whiteness* for an overview of Land's racism with regards to science fiction (which we would argue is its own kind of research-creation).
16. See Mosse's chapter titled "The Occult Origins of National Socialism" in *The Fascist Revolution* for an interesting analysis of the central role of occult practices and esoteric beliefs and its artists and theorists in the rise of Nazism.
17. A critical engagement with the occult practices and beliefs surrounding alchemy dates back to Hegel himself. See Glenn Alexander Magee's book *Hegel and the Hermetic Tradition.*
18. Warlick, *Max Ernst and Alchemy*, 103.
19. Benjamin, "Light from Obscurantists," 656.
20. This turn to the esoteric is, no doubt, the result of Benjamin's relationship with the noted scholar of Jewish mysticism Gershom Scholem. Indeed, Scholem even wrote a book titled *Alchemy and Kabbalah.*
21. Benjamin, "On Astrology," 685.
22. Benjamin, *Origin of the German Trauerspiel*, 10.
23. Benjamin, *Origin of the German Trauerspiel*, 10.
24. Principe, *The Secrets of Alchemy*, 68.
25. Benjamin, *Origin of the German Trauerspiel*, 27.
26. Benjamin, *The Arcades Project.*
27. In his book *Walter Benjamin and the Antinomies of Tradition*, McCole points to Benjamin's complex relationship with the occult (pre-fascist) writings of Ludwig Klages and how Benjamin appropriated Klages's notion of the *Urbilder*, transforming it into a historical rather than purely mythic image. On the opposite side of the spectrum, Benjamin also missed certain progressive potentials within surrealism's occult tendencies because he was too cautious and critical. As Walter Kalaidjian points out, Benjamin missed how feminist surrealists such as Leonora Carrington pushed the occult dimensions of surrealism, and in particular its alchemical leanings, toward emancipatory experiments in community, embodiment, and deep (dark) ecology. See Kalaidjian, "Occult Surrealism as 'Profane Illumination.'"
28. Letter, Benjamin to Scholem, August 8, 1937 (*GS*, 1:1070).
29. Löwy, *Fire Alarm*, 4.
30. Ramey's *The Hermetic Deleuze* is a definitive study of the impact of hermetic thought on Deleuze.
31. Deleuze and Guattari, *Thousand Plateaus*, 240.
32. Deleuze and Guattari, *Thousand Plateaus*, 240.
33. Deleuze and Guattari, *Thousand Plateaus*, 242.
34. Pignarre and Stengers, *Capitalist Sorcery*, 40.
35. Pignarre and Stengers, *Capitalist Sorcery*, 134.
36. Pignarre and Stengers, *Capitalist Sorcery*, 136.
37. Pignarre and Stengers, *Capitalist Sorcery*, 138.

38. Benjamin, "Paralipomena to 'On the Concept of History,'" 406.
39. Latour, *We Have Never Been Modern*, 69.
40. Latour, *We Have Never Been Modern*, 73.
41. Latour, *We Have Never Been Modern*, 75.
42. Bennett, *The Enchantment of Modern Life*.
43. Bennett, *Influx & Efflux*, 64.
44. According to Egil Asprem, the residual existence of the esoteric within the modern produces two phenomena: "the disenchantment of esoteric discourse on the one hand, and attempts to re-enchant science on the other" ("Psychic Enchantments of the Educated Classes: The Paranormal and the Ambiguities of Disenchantment," 334). It is precisely the scientific, political, and aesthetic ambiguities of this contact point that is the fulcrum of our analysis of the alchemical potentials of research-creation.
45. Benjamin, "Paralipomena to 'On the Concept of History,'" 406.
46. Benjamin, "Paralipomena to 'On the Concept of History,'" 405.
47. Benjamin, "Paralipomena to 'On the Concept of History,'" 397.
48. Egil Asprem and Kennet Granholm point out that such an approach to esoteric belief systems denies how traditions are constructed/invented by scholars and popular culture. This ahistorical notion of "authentic" tradition replays the logic of *philosophia perennis* or eternal wisdom that is active yet hidden within the present. Such beliefs are deeply conservative, and associated with Romantic notions of a folk spirit and national traditions uniting a people. See, Asprem and Granholm, "Constructing Esotericisms."
49. According to Hanegraaff in *Esotericism and the Academy*, the history of esoteric knowledge and occult practices in the West can be divided into three metahistorical narratives: evolution, degeneration, and continuity. To this list, we would like to add a fourth: transmutation through crypto-genealogy (a properly alchemical mode of historical mutation predicated on dark, unforeseen drifts).
50. Harney and Moten, *The Undercommons*.
51. Simon, "In Praise of the Paranormal Curiosity of Charles Fort, Patron Saint of Cranks."
52. Fort, *The Book of the Damned*, 15.
53. Fort, *The Book of the Damned*, 7.
54. Fort, *The Book of the Damned*, 25–26.
55. Agamben, "*Opus Alchymicum*."
56. For an extended and insightful overview of the role of the occult in Agamben's work (including astrology and alchemy) see Paul Colilli's *Agamben and the Signature of Astrology*.
57. Merkur, *Gnosis*, 74.
58. Atwood, *A Suggestive Inquiry into the Hermetic Mystery*, 143.
59. Jung, "The Idea of Redemption in Alchemy," 210.
60. Jarry, *Exploits & Opinions of Dr. Faustroll*, 104.
61. Brotchie, *Alfred Jarry*, 252.

62. Cited in Gorski, *Yeats and Alchemy*, 25–26.

63. Agamben, *The Fire and the Tale*, 137–38.

64. Agamben, *Autoritratto*, 91.

65. See the chapter titled "Fascism and the Intellectuals" in George Mosse's *The Fascist Revolution* for an overview of why fascism was so enticing for Yeats.

66. Melzer, *Philosophy Between the Lines*, 4.

67. Nummedal, "Sound and Vision," https://furnaceandfugue.org/essays/nummedal/.

68. Fulcanelli, *Dwellings of the Philosophers*, 57.

69. Heller-Roazen, *Dark Tongues*.

70. Toland, *Clidophorus*, 76.

71. Faivre, *Access to Western Esotericism*, 33.

72. Deleuze, "Mathesis, Science, and Philosophy," 143.

73. Deleuze, "Mathesis, Science, and Philosophy," 143.

74. Benjamin, "One-Way Street," 483.

75. Deleuze, *Difference and Repetition*, 119.

76. Cited in Eliade, *The Forge and the Crucible*, 164.

77. Vickers, "On the Function of Analogy in the Occult."

78. Principe, *Alchemy*, 111.

79. Roob, *Alchemy and Mysticism*, 76.

80. Benjamin, "On the Mimetic Faculty," 720.

81. Horkheimer and Adorno, *Dialectic of Enlightenment*, 57.

82. One can think here of Niklas Luhmann's theory of the different types of differentiation in Western modernity (including segmentation, stratification, and functional differentiation). See Luhmann, "Differentiation of Society."

83. Materer, *Modernist Alchemy*, 11.

84. Indeed, for Adorno and Horkheimer, mimesis in the modern era becomes so distorted that it turns into a mimesis of death supporting fascism. See *Dialectic of Enlightenment*, 44.

85. See Bennett, *Vibrant Matter*.

86. Marx, *Capital: Volume One*, 163–64.

87. Bennett, *Vibrant Matter*, xv.

88. Eliade, *The Forge and the Crucible*, 148.

89. Eliade, *The Forge and the Crucible*, 148.

90. Fulcanelli, *Dwellings of the Philosophers*, 29.

91. Here we have to part ways with Eliade's psychoanalytic interpretation of alchemy, which is viewed as the projection of human suffering and sacrifice onto matter through the death and rebirth of transmutation (151). This reading misses something more radical, which Eliade's history of alchemy highlights only to subsequently ignore in favor of a human-centric denial of the life of matter.

92. Faye, *Archeofuturism*, 58.

93. Faye, *Archeofuturism*, 72.

94. Faye, *Archeofuturism*, 173.

95. Materer, *Modernist Alchemy*, xiii.

96. Harney and Moten, *The Undercommons*.

97. Deleuze and Guattari, *Thousand Plateaus*, 252.

98. Deleuze and Guattari, *Thousand Plateaus*, 253.

99. Deleuze and Guattari, *Thousand Plateaus*, 253.

100. Cited in Principe, *Alchemy*, 44.

101. Melzer, *Philosophy Between the Lines*, 316.

102. Fort, *The Book of the Damned*, 17.

103. Fort, *The Book of the Damned*, 16.

104. Fort, *The Book of the Damned*, 17.

105. Freud, "Psycho-Analysis and Telepathy," 177–78.

106. Jones, *The Life and Work of Sigmund Freud*.

107. Materer, *Modernist Alchemy*, 21.

108. Lee, "Introduction: Perceptual Drift," 9.

109. Deleuze, *Difference and Repetition*, 144.

110. Manning, *Minor Gesture*, 14.

111. Manning, *Minor Gesture*, 112.

112. Yergeau, *Authoring Autism*, 111.

113. Here we can see a clear distinction between Adorno and Benjamin in relation to the occult. For Benjamin, his famous notion of the philosophical constellation remained connected to astrology, whereas for Adorno the very same concept had to be rigorously delineated from any connection to the occult in order for it to gain philosophical legitimacy. Although Adorno adopted the constellational methodology of thinking and writing from Benjamin, he had to disenchant it. Thus, in his inaugural academic lecture from 1931, Adorno wrote that philosophy has to "convey its elements into changing constellations—or, to use a less astrological and more scientifically current expression—into various experimental arrangements . . ." (cited in Mittelmeier, *Naples 1925*, 49).

Gate One

1. It is interesting to compare Davis's thesis with that proposed by Margaret Wertheim in her book *The Pearly Gates of Cyberspace* in which she convincingly argues that cyberspace is a technological proxy for the Christian space of heaven. Davis finds alchemical echoes in the infrastructure of digital, techno-culture whereas Wertheim turns to cyberspace as a substitute for heaven (despite the various optic fibers, microchips, and telecommunications satellites that comprise its all-too-earthly and physical infrastructure). Missed by Wertheim is therefore the dark, occult qualities underlying and supporting the fantasy of a pure, eternal, heavenly utopia.

2. Davis, *TechGnosis*, 53.

3. Eliade, *The Forge and the Crucible*, 172–73.

4. Two exceptions to this trend are the work of Anirban Gupta-Nigam on "black infrastructure" and Jeffrey Di Leo's analysis of "dark

infrastructures" of the "dark academe." Di Leo's chapter and our own can be read as two faces of a dialectic of dark infrastructure, thus revealing the dangers and potentialities of the occult dimensions of such invisible forces, materials, and networks.
5. Larkin, "The Politics and Poetics of Infrastructure," 329.
6. Larkin, "The Politics and Poetics of Infrastructure," 336.
7. Deleuze and Guattari, *Anti-Oedipus: Capitalism and Schizophrenia*, 348.
8. Deleuze and Guattari, *What Is Philosophy?*, 41.
9. Berlant, "The Commons: Infrastructures for Troubling Times," 403.
10. Berlant, "The Commons: Infrastructures for Troubling Times," 404.
11. Berlant, "The Commons: Infrastructures for Troubling Times," 406.
12. Berlant, "The Commons: Infrastructures for Troubling Times," 407.
13. The idea of an economic base or *Grundlage* has also been translated into French as *infrastructure*. Yet the "economic structure of society" or "real basis" or "real foundation" is, at least in classical Marxist thought different from the English understanding of infrastructure in two important respects. First, economic basis is much narrower than infrastructure, and second, the power of determination attributed to this economic basis is lacking in discussions of infrastructure (indeed, linguistic, cultural, and technical infrastructure is often interpreted as determined by a preceding and controlling economic base). As such, the two overlap but are not exactly synonymous.
14. Kishik, *The Manhattan Project*, 75.
15. Kishik, *The Manhattan Project*, 76.
16. Easterling, *The Action Is the Form*, 7.
17. Easterling, *The Action Is the Form*, 30.
18. Easterling, *The Action Is the Form*, 43.
19. Paglen, *Blank Spots on the Map*.
20. Galloway, "What Is Hermeneutic Light?"
21. Deleuze, *Negotiations, 1972–1990*, 120.
22. Laruelle, "On the Black Universe," 105.
23. Michael Rubenstein, Bruce Robbins, and Sophia Beal, "Infrastructuralism: An Introduction."
24. Christopher Breu and Jeffrey R. Di Leo (2023), "Theorizing Infrastructure: An Introduction."
25. North, *The Yield*, 191.
26. Moten, "The Case of Blackness," 187.
27. Moten, "The Case of Blackness," 202.
28. Culp, *A Guerrilla Guide to Refusal*, 133.
29. Browne, *Dark Matters*.
30. Bryant, "Dark Objects."
31. Michel Serres, *The Five Senses*, 67.
32. Glissant, *Poetics of Relation*, 65.
33. Dufourmantelle, *In Praise of Risk*, 168.
34. Thacker, *Starry Speculative Corpse*, 55.
35. Thacker, *Starry Speculative Corpse*, 55.

36. Culp, *A Guerrilla Guide to Refusal*, 8.
37. Culp, *A Guerrilla Guide to Refusal*, 22.
38. Culp, *A Guerrilla Guide to Refusal*, 42.
39. Culp, *A Guerrilla Guide to Refusal*, 43.
40. Dixon-Román and Amaro, "Haunting, Blackness, and Algorithmic Thought."
41. Critical Computation Bureau, "Editorial."
42. Lewis, *On Study*.
43. Ford and Lewis, "On the Freedom to Be Opaque Monsters."
44. Harney and Moten, *The Undercommons*.
45. As emphasized by Di Leo in his book *Dark Academe* and in particular the chapter on dark infrastructure.
46. Morton, *Dark Ecology*, 7.
47. Morton, *Dark Ecology*, 42.
48. Morton, *Dark Ecology*, 42.
49. As described by Brian O'Doherty who argues that the modern studio has, in many instances, crossed over into the gallery space, *Studio and Cube*.
50. Le Corbusier, "A Coat of Whitewash," 192.
51. Caillois, *Edge of Surrealism*, 98–99.
52. Caillois, *Edge of Surrealism*, 98.
53. Dufourmantelle, *In Praise of Risk*, 173.
54. Dufourmantelle, *In Praise of Risk*, 175.
55. Hall, *The Artist's Studio*, 73.
56. Hall, *The Artist's Studio*, 126.
57. Pallasmaa, *The Eyes of the Skin*, 39.
58. Hall, *The Artist's Studio*, 126.
59. "Studio Tour with Betye Saar."
60. "Studio Tour with Betye Saar."
61. "Studio Tour with Betye Saar."
62. "Betye Saar: The Legends of Black Girl's Window."
63. O'Doherty, *Studio and Cube*, 18.
64. O'Doherty, *Studio and Cube*, 18.
65. "Betye Saar: The Legends of Black Girl's Window."
66. Quoted in Sills, *Visions*, 38.
67. Lee, "Introduction: Defining Perceptual Drift," 9.
68. Deleuze and Guattari, *Thousand Plateaus*, 240.
69. Friedberg, *The Virtual Window*, 2–3.
70. wilson, Acuff, and López, "Conjuring Hands: The Art of Curious Women of Color."

Gate Two

1. Mitchell, *Experimental Life*, 21.
2. Mitchell, *Experimental Life*, 41.
3. Daston, *Rules*, 265.

4. Daston, *Rules*, 36.

5. Daston, *Rules*, 272.

6. Daston, *Rules*, 7.

7. Daston and Galison, "The Image of Objectivity."

8. Harney and Moten, *Undercommons*, 92.

9. Principe, *Alchemy*, 121.

10. Newman and Principe, *Alchemy. Tried in Fire.*

11. Newman and Principe, 57.

12. Price, "Introduction," 7.

13. Price, "Introduction," 6.

14. Jarry, *Faustroll*, 21.

15. Jarry, *Faustroll*, 22.

16. Bök, *Pataphysics*, 9.

17. Bök, *Pataphysics*, 70.

18. Bök, *Pataphysics*, 79.

19. Bök, *Pataphysics*, 79.

20. Jarry, *How to Construct a Time Machine*, 5.

21. Latour and Woolgar, *Laboratory Life*, 83.

22. Latour and Woolgar, *Laboratory Life*, 240.

23. Latour and Woolgar, *Laboratory Life*, 244–45.

24. Latour and Woolgar, *Laboratory Life*, 244.

25. An even more dramatic description of the inoperative laboratory can be found in the speculative fiction titled "Nonhuman Findings from the Laboratory of Speculative Sociology" by Elizabeth de Freitas in which she transforms a rather mundane laboratory setting into an alchemical studio for the germination of vestigial sensibilities between art and science.

26. Deleuze and Guattari, *Anti-Oedipus*, 78.

27. Deleuze and Guattari, *Anti-Oedipus*, 84.

28. Deleuze and Guattari, *Anti-Oedipus*, 7.

29. Deleuze and Guattari, *Thousand Plateaus*, 373.

30. Deleuze and Guattari, *Thousand Plateaus*, 373.

31. Deleuze and Guattari, *Thousand Plateaus*, 373.

32. Deleuze and Guattari, *Thousand Plateaus*, 373.

33. Deleuze and Guattari, *Thousand Plateaus*, 373.

34. Agamben, *The Use of Bodies*, 229. In Italian, Agamben utilizes the word *erranza*, which means to wander, but we appreciate the translator, Adam Kotsko's, more intuitive notion of drift. For us, drift is more than wandering, as it can be a verb and a noun, indicating an accumulation or clump. This dialectical notion of drift captures Agamben's sense of a life (a loose, frayed, and never fully distinct pattern) that emerges from inside the wanderings of children—the paradigm of which are autistic children.

35. Deleuze and Guattari, *Thousand Plateaus*, 14.

36. Agamben, *The Use of Bodies*, 241.

37. Manning, *For a Pragmatics of the Useless*, 72.

38. Manning, *For a Pragmatics of the Useless*, 72.

39. Manning, *For a Pragmatics of the Useless*, 72.
40. Manning, *For a Pragmatics of the Useless*, 70.
41. Manning, *Minor Gesture*, 117.
42. Manning, *Minor Gesture*, 216.
43. Manning, *Minor Gesture*, 225.
44. Manning, *Minor Gesture*, 59.
45. Berlant, "The Commons: Infrastructures for Troubling Times," 394.
46. These distinctions between flux and drift and the distinction between flow and clumping within drift are our humble contributions to a number of philosophies that take up the question of flow (in one way or another) including Manning in *For a Pragmatics of the Useless*, Michel Serres in *The Five Senses*, and Jane Bennett in *Influx & Efflux* (to name a few).
47. Manning, *For a Pragmatics of the Useless*, 76.
48. Harney and Moten, *Undercommons*, 97.
49. Han, *Scent of Time*, 30.
50. Han, *Scent of Time*, 31.
51. Han, *Scent of Time*, 39.
52. Han, *Scent of Time*, 41.
53. Han, *Scent of Time*, 93.
54. Han, *Scent of Time*, 93.
55. Han, *Scent of Time*, 67.
56. McCole, *Walter Benjamin and the Antinomies of Tradition*.
57. Agamben, *Autoritratto nello Studio*, 69.
58. Natalie Loveless and Erin Manning, "Research-Creation as Interdisciplinary Praxis," 213.
59. Hall, *The Artist's Studio*, 74.
60. Hofstadter, *I Am a Strange Loop*, 101–102.
61. Jarry, *Faustroll*, 99.
62. Morton, *Dark Ecology*, 7.
63. One exception to this is the chapter "From No-ing to Knowing, from Naughts to Knots" by Paul Couillard in Loveless's collection. Yet even Couillard does not see how knotting is an inherent aspect of the distinct drift of research-creation as an alchemical and pataphysical form of studioing.
64. Deleuze and Guattari, *Thousand Plateaus*, 311–12.
65. Principe, *Alchemy*, 203.
66. Deleuze, *Foucault*, 34.
67. Deleuze, *Foucault*, 35.
68. Deleuze, *Foucault*, 34.
69. Deleuze, *Foucault*, 44.
70. Deleuze, *Foucault*, 43.
71. Haraway, *Staying with the Trouble*, 79.
72. Haraway, *Staying with the Trouble*, 4.
73. Haraway, *Staying with the Trouble*, 12.
74. Haraway, *Staying with the Trouble*, 33.

75. Haraway, *Staying with the Trouble*, 31.
76. Manning, *Minor Gesture*, 67.
77. Manning, *Minor Gesture*, 67.
78. Pignarre and Stengers, *Capitalist Sorcery*, 140.
79. For an analysis of the logic of learning vs. the logic of studioing see Hyland and Lewis, *Studious Drift*.
80. Kahn and Lewis, *Education out of Bounds*.

Gate Three

1. Dufourmantelle, *In Praise of Risk*, 2.
2. Serres, *Five Senses*, 29.
3. Serres, *Five Senses*, 29.
4. Manning, *Minor Gesture*, 53.
5. Manning, *Minor Gesture*, 218.
6. Jackson, "Analysis and Synthesis in Nineteenth-Century Organic Chemistry."
7. In this sense, our chapter is a furtherance of Erin Manning's attempt to distinguish the act of creation possible in research-creation from the creativity of capitalism (and for us, the dialectical inversion of such creativity in the form of fascist nihilism). See Sarah E. Truman, Natalie Loveless, Erin Manning, Natasha Myers, and Stephanie Springgay, "The Intimacies of Doing Research-Creation," 228.
8. Agamben, *Profanations*, 81.
9. Agamben, *The Sacrament of Language*, 31.
10. Agamben, *Profanations*, 75.
11. Agamben, *Profanations*, 77.
12. Kotsko, *Awkwardness*.
13. Kotsko, *Awkwardness*, 65.
14. Kotsko, *Awkwardness*, 69.
15. Kotsko, *Awkwardness*, 79.
16. Manning, *Minor Gesture*, 24.
17. Deleuze, *Negotiations*, 158.
18. Braidotti, *Transpositions*, 217.
19. Carrington, *Down Below*, 22.
20. Carrington, *Down Below*, 46.
21. Carrington, *Down Below*, 46.
22. Carrington, *Down Below*, 46.
23. Carrington, *Down Below*, 46–47.
24. Carrington, *Down Below*, 47.
25. Carrington, *Down Below*, 49.
26. Carrington, *Down Below*, 56.
27. Deleuze and Guattari, *A Thousand Plateaus*, 503.
28. Dufourmantelle, *In Praise of Risk*, 60.

29. Dufourmantelle, *In Praise of Risk*, 60.
30. Dufourmantelle, *In Praise of Risk*, 60.
31. Franklin, *The Cult of Creativity*.
32. Benjamin, "The Work of Art in the Age of Its Reproducibility," 104–105.
33. Dufourmantelle, *In Praise of Risk*, 60.
34. Deleuze and Guattari, *Thousand Plateaus*, 163.
35. Deleuze and Guattari, *Thousand Plateaus*, 160.

Gate Four

1. Lewis, *Walter Benjamin's Antifascist Education*.
2. Connolly, *Aspirational Fascism*.
3. Toscano, *Late Fascism*.
4. Antliff, *Avant-Garde Fascism*.
5. Benjamin, *Selected Writings, Volume 3: 1935–1938*, 129.
6. Benjamin, *Selected Writings, Volume 3: 1935–1938*, 129.
7. Benjamin, *Selected Writings, Volume 3: 1935–1938*, 129.
8. Benjamin, *Selected Writings, Volume 3: 1935–1938*, 129.
9. See Adorno, "Education After Auschwitz," in *Critical Models*.
10. Klemperer, *The Language of the Third Reich*, 15–16.
11. Benjamin, *Selected Writings, Volume 3: 1935–1938*, 129.
12. Bataille, "The Structure of Fascism," 122.
13. Bataille, "The Structure of Fascism," 122–23.
14. Bataille, "The Structure of Fascism," 126.
15. Bataille, "The Structure of Fascism," 126.
16. Bataille, "The Structure of Fascism," 127.
17. Bataille, "The Structure of Fascism," 128.
18. Bataille, "The Structure of Fascism," 128–29.
19. Bataille, "The Structure of Fascism," 132.
20. Bataille, "The Structure of Fascism," 142.
21. See for instance, Wolin, "Left Fascism."
22. Bataille, "The Structure of Fascism," 135.
23. Bataille, "Toward Real Revolution," 37.
24. Bataille, "Toward Real Revolution," 145.
25. Blanchot, *The Unavowable Community*, 7.
26. Preface by Foucault in *Anti-Oedipus* by Deleuze and Guattari, xiii.
27. Preface by Foucault in *Anti-Oedipus* by Deleuze and Guattari, xiii.
28. Deleuze and Guattari, *Anti-Oedipus*, 232.
29. Deleuze and Guattari, *Anti-Oedipus*, 35.
30. Deleuze and Guattari, *Anti-Oedipus*, 258.
31. Deleuze and Guattari, *Anti-Oedipus*, 340.
32. Deleuze and Guattari, *Anti-Oedipus*, 340.
33. Deleuze and Guattari, *Thousand Plateaus*, 34.
34. Deleuze and Guattari, *Thousand Plateaus*, 34.

35. Deleuze and Guattari, *Thousand Plateaus*, 342.
36. Deleuze and Guattari, *Thousand Plateaus*, 340.
37. Deleuze and Guattari, *Thousand Plateaus*, 309.
38. Guattari, *Chaosophy*.
39. Guattari, *Chaosophy*, 159–60.
40. Guattari, *Chaosophy*.
41. Deleuze and Guattari, *Anti-Oedipus*, 239.
42. For instance, Joseph Richard Winters convincingly connects Moten's theory of blackness with Bataille's complex notion of the heterogenous in his essay titled "Between Ecstasy and Abjection: Black Studies and the Excessive Sacred."
43. Moten, "The Case of Blackness," 187.
44. Moten, "The Case of Blackness," 202.
45. Moten, "Blackness and Nothingness (Mysticism in the Flesh)," 740.
46. Moten, "Blackness and Nothingness (Mysticism in the Flesh)," 760.
47. Moten, "Blackness and Nothingness (Mysticism in the Flesh)," 756.
48. Moten, "Blackness and Nothingness (Mysticism in the Flesh)," 742.
49. Moten, "Blackness and Nothingness (Mysticism in the Flesh)," 742.
50. Moten, "Blackness and Nothingness (Mysticism in the Flesh)," 753.
51. Moten, "Blackness and Nothingness (Mysticism in the Flesh)," 756.
52. Moten, "Blackness and Nothingness (Mysticism in the Flesh)," 761.
53. Moten, "Blackness and Nothingness (Mysticism in the Flesh)," 761.
54. Bacon, *Radix Mundi*, in *The Alchemy Reader*, 117.
55. Bacon, *Radix Mundi*, in *The Alchemy Reader*, 118.
56. See Corbin, *The Man of Light in Iranian Sufism* for an overview of alchemical optics.
57. Mosse, *The Fascist Revolution*, 100.
58. Moten, "Blackness and Nothingness (Mysticism in the Flesh)," 765.
59. Harney and Moten, "Plantocracy or Communism," 53.
60. Harney and Moten, "Plantocracy or Communism," 62.
61. Harney and Moten, "Plantocracy or Communism," 63.
62. Harney and Moten, "Plantocracy or Communism," 62.
63. Lewis, "The Pataphysics of Inoperativity in the Works of Giorgio Agamben."
64. Deleuze and Guattari, *Thousand Plateaus*, 25.
65. Agamben, *State of Exception*.
66. Benjamin, "On the Concept of History," 392.
67. Agamben, *State of Exception*.
68. Grindon, "Alchemist of the Revolution," 305–17.
69. Deleuze, "The Exhausted."
70. Deleuze and Guattari, *Thousand Plateaus*, 24.
71. McCaffery, *The Darkness of the Present*, 183.
72. McCaffery, *The Darkness of the Present*, 184.
73. Toscano, *Late Fascism*.
74. See Forshaw, *"Oratorium—Auditorium—Laboratorium."*
75. Anderson and Jones, "Introduction: The Rise of Astro-Blackness," vii.

76. Faye, *Archeofuturism*.
77. Cited in Gaskins, "Afrofuturism on Web 3.0: Vernacular Cartography and Augmented Space," 31.
78. Manning on blackness and neurodiversity in *The Minor Gesture*.
79. Eshun, *More Brilliant Than the Sun*, 162.
80. Worby, "Cacophony," 161.
81. Moten, "Blackness and Nothingness (Mysticism in the Flesh)," 774.
82. Adorno, "On the Fetish Character in Music and the Regression of Listening."
83. Nabeel Zuberi, "The Transmolecularization of [Black] Folk," 92.
84. Rediker, *The Slave Ship*, 307.
85. Rediker, *The Slave Ship*, 278.
86. van Veen, "The Armageddon Effect," 74.
87. In this sense, we somewhat disagree with Kodwo Eshun's reading of Sun Ra as a reincarnated despot. See Eshun's *More Brilliant Than the Sun*.
88. Eshun, *More Brilliant Than the Sun*, 154.
89. van Veen, "The Armageddon Effect."
90. It should be noted that the contemporary digital artist and utopian Afrofuturist Cauleen Smith has argued that "We—all of us—are without a home in space-time. And I am suggesting that is how it should be" ("We Already Have What We Need"). This position challenges the idea of a homeland, a concept easily co-opted by colonialist projects. Appropriating Smith's insight, we might argue that she does not negate Sun Ra but rather completes his gesture. Space as the place of drift *becomes a pure means* rather than a mere transitional space on the way to a final destination, another homeland.

Gate Five

1. Cramer, "What Is 'Post-Digital'?"
2. Galloway, *Uncomputable*.
3. See Daston, *Rules*.
4. Tavin, Kallio-Tavin, and Ryynänen, *Art, Excess, and Education*.
5. Hugill and Hendler, "Pataphysics and Computing," in *Pataphysics Unrolled*, 259.
6. Johanna Drucker, "The Patacritical Demon, Two Decades Later," in *Pataphysics Unrolled*, 197.
7. Hugill and Hendler, "Pataphysics and Computing," in *Pataphysics Unrolled*, 276, 277.
8. Ritchie, "The Demon in the Diagram," 217.
9. The themes of non-knowledge against the transparency and calculability of knowledge in relation to anti-capitalist, postdigital research has been introduced in the work of Ford, Swenson, and Fosher (2022). The present gate is, in many ways, a restaging of these themes with regards to Bataille.

10. Cronenberg, *Crimes of the Future*.
11. Jarry, *Exploits & Opinions of Dr. Faustroll, Pataphysician*, 21.
12. It is important to note that even if Bataille does not cite pataphysics per se, Jarry's aesthetic concerns with the occult, ritual, and pornography certainly contributed to Bataille's explorations into tabooed erotica, not to mention their shared concern with sovereign power. Furthermore, Bataille's friend and collaborator Raymond Queneu was a well-known pataphysical writer.
13. Bataille, *The Sacred Conspiracy*, 334.
14. Bataille, *The Sacred Conspiracy*, 334.
15. Bataille, *The Bataille Reader*, 169.
16. Bataille, *The Bataille Reader*, 176.
17. There is some tension here between our earlier notion of the profane and Bataille's notion of the sacred. For simplicity, we opt to think of Bataille's unusual notion of the sacred as a profanation of the sacred, or as an inoperative notion of the sacred (as it is commonly understood).
18. While Agamben ultimately rejected Bataille's politics as neofascist (even if Bataille resolutely stood against all forms of political, social, and military servility to fascist dictators), it is unquestionable that Agamben's treatment of sovereign power, states of exception, the sacred, and so forth are in dialogue with both Jarry and Bataille.
19. Bataille, *The Sacred Conspiracy*, 330.
20. Bataille, *The Sacred Conspiracy*, 330.
21. Bruno, "Les techniques d'illumination chez Georges Bataille."
22. Interestingly, Bataille defines his religion as nothing more than the "act of laughing (or tears, or erotic stimulation) in the precise sense that laughter (like tears or erotic stimulation) represents the defeat of everything that had sought to impose its permanence" (*The Sacred Conspiracy*, 326).
23. Bök, *Pataphysics*, 64.
24. Bénabou, "Rules and Constraint," 41.
25. Bénabou, "Rules and Constraint," 41.
26. Calvino, "Prose and Anticombinatorics."
27. Bataille, *The Unfinished System of Nonknowledge*, 133.
28. Bataille, *The Unfinished System of Nonknowledge*, 133.
29. Bataille, *The Unfinished System of Nonknowledge*, 129.
30. Bataille, *The Unfinished System of Nonknowledge*, 131.
31. Bataille, *The Unfinished System of Nonknowledge*, 5.
32. Bataille, *The Unfinished System of Nonknowledge*, 16.
33. Bataille, *The Unfinished System of Nonknowledge*, 16.
34. Jarry, *Exploits & Opinions of Dr. Faustroll, Pataphysician*, 22.
35. Merleau-Ponty, *Phenomenology of Perception*, 248.
36. Serres, *The Five Senses*, 26.
37. See Haraway, "Situated Knowledges."
38. Bataille, *The Sacred Conspiracy*, 125.
39. Bataille, *The Sacred Conspiracy*, 125.
40. Ahmed, *What's the Use?*, 219.

41. Ahmed, *What's the Use?*, 219.
42. Bataille, *The Sacred Conspiracy*, 125.
43. Garoian, *The Prosthetic Pedagogy of Art*.
44. Garoian, *The Prosthetic Pedagogy of Art*, 15.
45. Garoian, *The Prosthetic Pedagogy of Art*, 18–19.
46. Bataille, *The Unfinished System of Nonknowledge*, 137.
47. Bataille, *The Unfinished System of Nonknowledge*, 136.
48. Yergeau, *Authoring Autism*, 112.
49. Beard, *The Artist as Monster*, 16.
50. Smith, "DisORIENTations: Travels Through Blackness."

Vestigial Comments

1. Agamben, *Pinocchio*, 8.
2. Cited in Eliade, *The Forge and the Crucible*, 163–64.
3. Agamben, *Infancy and History*, 94.
4. Here we take inspiration from Rosemarie Garland-Thomson. Using Garland-Thomson's analysis of misfitting, our book becomes a "concatenation of misfits" composed of "awkward encounters, graceless engagements between things in the material world that end in an incongruent relationship" ("What Misfitting Makes," 16). The awkwardness of such encounters is generative, leading to new uses, new tastes, new flavors of becoming. There is an alchemy that is only possible when misfits assemble through *discordia concors*.
5. Manning, *For a Pragmatics of the Useless*, 75.
6. Manning, *For a Pragmatics of the Useless*, 76.
7. Ahmed, *What's the Use?*, 219.
8. Vidler, *The Architectural Uncanny*, 168.
9. Vidler, *The Architectural Uncanny*, 172.
10. Deleuze and Guattari, *Thousand Plateaus*, 253.
11. Kafka, "The Cares of a Family Man," 428.
12. Kafka, "The Cares of a Family Man," 428.
13. On corners in Kafka's work see North, *The Yield*, 191. Indeed, we would argue that the corner is the essential aspect of dark infrastructure in Kafka's work.
14. Agamben, *Potentialities*, 223.
15. Benjamin, *Selected Writings Volume One*, 449–50.
16. Agamben, *The Fire and the Tale*, 137–38.
17. For an excellent art historical and philosophical overview of this work in relation to Varo's larger *opus*, see the collection of essays in *Remedios Varo: Science Fictions*, edited by Cailtin Haskell and Tere Arcq.
18. Serres, *The Five Senses*, 146.
19. Serres, *The Five Senses*, 128.
20. Serres, *The Five Senses*, 112–13.

Bibliography

Aberth, Susan L. *Leonora Carrington: Surrealism, Alchemy and Art*. Lund Humphries, 2010.

Adorno, Theodor W. *Critical Models: Interventions and Catchwords*. Translated by Henry W. Rickford. Columbia University Press, 1998.

Adorno, Theodor W. "Freudian Theory and the Pattern of Fascist Propaganda." In *The Essential Frankfurt School Reader*, edited by Andrew Arato and Eike Gebhardt. Continuum, 2002.

Adorno, Theodor W. *Minima Moralia: Reflections from Damaged Life*. Translated by E. F. N. Jephcott. Verso Books, 2005.

Adorno, Theodor W. "On the Fetish Character in Music and the Regression of Listening." In *The Culture Industry*, edited by Jay M. Bernstein. Routledge, 1991.

Adorno, Theodor W. "Types and Syndromes." In *The Authoritarian Personality*, edited by Theodor W. Adorno, Else Frenkel-Brunswik, Daniel J. Levinson, and R. Nevitt Sandord. Harber & Brothers, 1950.

Agamben, Giorgio. *Autoritratto nello Studio*. Luce Mediterranea, 2017.

Agamben, Giorgio. "Bataille and the Paradox of Sovereignty." Translated by Michael Krimper. *Journal of Italian Philosophy* 3, (2020a): 247–53.

Agamben, Giorgio. *Homo Sacer: Sovereign Power and Bare Life*. Translated by Daniel Heller-Roazen. Stanford University Press, 1999.

Agamben, Giorgio. *Infancy and History: On the Destruction of Experience*. Translated by Liz Heron. Verso, 2007.

Agamben, Giorgio. "On Potentiality." In *Potentialities: Collected Essays in Philosophy*, translated by Daniel Heller-Roazen. Stanford University Press, 2000.

Agamben, Giorgio. "Opus Alchymicum." In *The Fire and the Tale*, translated by Lorenzo Chiesa. Stanford University Press, 2017.

Agamben, Giorgio. *Pinocchio: The Adventures of a Puppet, Doubly Commented Upon and Triply Illustrated*. Translated by Adam Kotsko. Seagull Books, 2023.

Agamben, Giorgio. *Profanations*. Translated by Jeff Fort. Zone Books, 2020.

Agamben, Giorgio. *The Sacrament of Language: An Archaeology of the Oath*. Translated by Adam Kotsko. Stanford University Press, 2011.

Agamben, Giorgio. *Stanzas: Word and Phantasm in Western Culture*. Translated by Ronald L. Martinez. University of Minnesota Press, 1993.

Agamben, Giorgio. *State of Exception*. Translated by Kevin Attell. University of Chicago Press, 2005.

Agamben, Giorgio. *Taste*. Translated by Cooper Francis. Seagull Books, 2017.

Agamben, Giorgio. *The Use of Bodies*. Translated by Adam Kotsko. Stanford University Press, 2016.

Ahmed, Sara. *Queer Phenomenology: Orientations, Objects, Others*. Duke University Press, 2006.

Ahmed, Sara. *What's the Use? On the Uses of Bodies*. Duke University Press, 2019.

Algazi, Gadi. "At the Study: Notes on the Production of the Scholarly Self." In *Space and Self in Early Modern European Cultures*, edited by D. W. Sabean and M. Stefanovska. University of Toronto Press, 2012.

Anderson, Reynaldo, and Charles E. Jones. "Introduction: The Rise of Astro-Blackness." In *Afrofuturism 2.0: The Rise of Astro-Blackness*. Lexington Books, 2015.

Antliff, Mark. *Avant-Garde Fascism: The Mobilization of Myth, Art, and Culture in France, 1909–1939*. Duke University Press, 2007.

Arendt, Hannah. *The Human Condition*, 2nd ed., University of Chicago Press, 2018.

Asprem, Egil. "Psychic Enchantments of the Educated Classes: The Paranormal and the Ambiguities of Disenchantment." In *Contemporary Esotericism*, edited by Egil Asprem and Kennet Granholm. Routledge, 2014.

Asprem, Egil, and Kennet Granholm. "Constructing Esotericisms: Sociological, Historical and Critical Approaches to the Invention of Tradition." In *Contemporary Esotericism*, edited by Egil Asprem and Kennet Granholm. Routledge, 2014.

Atwood, Mary Anne. *A Suggestive Inquiry into the Hermetic Mystery*. Trelawney Sunders Charing Cross, 1850.

Bacon, Roger. "From the *Radix Mundi*." In *The Alchemy Reader*, edited by Stanton J. Linden. Cambridge University Press, 2003.

Bacon, Roger. *The Mirror of Alchimy*. Richard Olive, 1597.

Bataille, Georges. "The Structure of Fascism." In *The Bataille Reader*, edited by Fred Botting and Scott Wilson. Routledge, 1997.

Bataille, Georges. *Encyclopedia Acephalica*. Edited by Robert Lebel and Isabelle Waldberg. Atlas Press, 1995.

Bataille, Georges. "The Notion of Expenditure." In *Visions of Excess: Selected Writings, 1927–1939*, translated and edited by Allan Stoekl. University of Minnesota Press 1984.

Bataille, Georges. *The Sacred Conspiracy: The Internal Papers of the Secret Society of Acéphale and Lectures to the College of Sociology*. Edited by Marina Galletti and Alestair Brotchie. Atlas Press, 2017.

Bataille, Georges. "Toward Real Revolution." Translated by Anette Michelson. *October* 36 (1986): 32–41. https://doi.org/10.2307/778547.

Bataille, Georges. *The Unfinished System of Nonknowledge*. Edited by Stuart Kendall. Translated by Michelle Kendall and Stuart Kendall. University of Minnesota Press, 2001.

Beard, William. *The Artist as Monster: The Cinema of David Cronenberg*. University of Toronto Press, 2006.

Bénabou, Marcel. "Rules and Constraint." In *Oulipo: A Primer of Potential Literature*, edited by Warren F. Motte. Dalkey Archive Press, 1986.

Benjamin, Walter. *The Arcades Project*. Belknap Press of Harvard University Press, 1999.

Benjamin, Walter. *Gesammelte Schriften.* 7 vols. Frankfurt am Main: Suhrkamp, 1991.

Benjamin, Walter. "Light from Obscurantists." In *Selected Writings Volume 2, Part 2 1931–1934*, edited by Michael W. Jennings, Howard Eiland, and Gary Smith. Harvard University Press, 2005.

Benjamin, Walter. "On Astrology." In *Selected Writings Volume 2, Part 2 1931–1934*, edited by Michael W. Jennings, Howard Eiland, and Gary Smith. Harvard University Press, 2005.

Benjamin, Walter. "One-Way Street." In *Selected Writings Volume 1, 1913–1926*, edited by Marcus Bullock and Michael W. Jennings. Harvard University Press, 2005.

Benjamin, Walter. "On the Concept of History." In *Selected Writings Volume 4, 1938–1940*, edited by Howard Eiland and Michael W. Jennings. Harvard University Press, 1999.

Benjamin, Walter. "On the Mimetic Faculty." In *Selected Writings Volume 2, Part 2 1931–1934*, edited by Michael W. Jennings, Howard Eiland, and Gary Smith. Harvard University Press, 2005.

Benjamin, Walter. *Origin of the German Trauerspiel.* Translated by Howard Eiland. Harvard University Press, 2019.

Benjamin, Walter. "Paralipomena to 'On the Concept of History.'" In *Selected Writings Volume 4, 1938–1940*, edited by Howard Eiland and Michael W. Jennings. Harvard University Press, 1999.

Benjamin, Walter. "The Work of Art in the Age of Mechanical Reproduction." In *Selected Writings Volume 3, 1935–1938*, edited by Howard Eiland and Michael W. Jennings. Harvard University Press, 2004.

Bennett, Jane. *The Enchantment of Modern Life: Attachments, Crossings, and Ethics.* Princeton University Press, 2001.

Bennett, Jane. *Vibrant Matter: A Political Ecology of Things.* Duke University Press, 2010.

Bennett, Jane. *Influx & Efflux: Writing Up with Walt Whitman.* Duke University Press, 2020.

Berlant, Lauren. "The Commons: Infrastructures for Troubling Times." *Environment Planning D: Society and Space* 34, no. 3 (2016): 393–419.

"Betye Saar: The Legends of Black Girl's Window," MoMA exhibition video. https://www.youtube.com/watch?v=G2iK2LHttfc.

Bishop, Claire. "Information Overload." *Art Forum* 2023 https://www.artforum.com/features/claire-bishop-on-the-superabundance-of-research-based-art-252571/.

Blanchot, Maurice. *The Unavowable Community.* Station Hill Press, 2006.

Bök, Christian. *Pataphysics: The Poetics of an Imaginary Science.* Northwestern University Press, 2002.

Braidotti, Rosi. *Transpositions: On Nomadic Ethics.* Translated by Michael Heyns. Polity Press, 2006.

Breu, Christopher, and Jeffrey R. Di Leo. "Theorizing Infrastructure: An Introduction." *Symplokē* 31, no. 1–2 (2023): 1–8. https://doi.org/10.1353/sym.2023.a914647.

Brinkhurst-Cuff, Charlie. "Turbine Hall Artist Kara Walker: 'Apparently, the Only Thing I Am Is Black.'" *The Guardian*, September 23, 2019. https://www.theguardian.com/artanddesign/2019/sep/23/kara-walker-turbine-hall-tate-modern-racially-charged.

Brotchie, Alastair. *Alfred Jarry: A Pataphysical Life*. MIT Press, 2015.

Browne, Simone. *Dark Matters: On the Surveillance of Blackness*. Duke University Press, 2015.

Bruno, Jean. "Les techniques d'illumination chez Georges Bataille." *Critique* 19 (1963): 195–96.

Bryant, Levi. "Dark Objects." *Larval Subjects*, May 25, 2011. https://larvalsubjects.wordpress.com/2011/05/25/dark-objects/.

Burckhardt, Titus. *Alchemy: Science of the Cosmos, Science of the Soul*. Translated by William Stoddart. Penguin Books, 1972.

Caillois, Roger. *The Edge of Surrealism: A Roger Caillois Reader*. Edited by Claudine Frank. Duke University Press, 2003.

Calvino, Italio. "Prose and Anticombinatorics." In *Oulipo: A Primer of Potential Literature*, edited and translated by Warren F. Motte. Dalkey Archive Press, 1986.

Carelman, Jacques. *A Catalogue of Unfindable Objects: Objets Introuvables*. Translated by Michael Bayley. Century Hutchinson, 1984.

Carrington, Leonora. *Down Below*. Edited by Marina Warner. New York Review Books, 2017.

Carroll, Jordan S. *Speculative Whiteness: Science Fiction and the Alt-Right*. University of Minnesota Press, 2024.

Cennini, Cennino. *The Book of the Art of Cennino Cennini: A Contemporary Practical Treatise on Quattrocento Painting*. Translated by Christiana J. Herringham. Routledge, 2018.

Chaucer, Geoffrey. "The Canon's Yeoman's Tale." In *The Canterbury Tales*, edited and translated by Nevill Coghill. Penguin Classics, 2003.

Checker, Chubby. "Let's Twist Again." Track #19 on *It's Pony Time/Let's Twist Again*, 7004–LP, P 7004, 1961, LP.

Chung, Sheng Kuan. "Media/Visual Literacy Art Education: Sexism in Hip-Hop Music Videos." *Art Education* 60, no. 3 (2007): 33–38.

Chung, Sheng Kuan, and Michael S. Kirby. "Media Literacy Art Education: Logos, Culture, Hamming, and Activism." *Art Education* 62, no. 1 (2009): 34–39.

Citarella, Joshua. *Politigram and the Post-left*. Blurb, 2024.

Cole, Michael W., and Mary Pardo. "Origin of the Studio." In *Inventions of the Studio, Renaissance to Romanticism*, edited by Michael W. Cole and Mary Pardo. University of North Carolina Press, 2005.

Colilli, Paul. *Agamben and the Signature of Astrology: Spheres of Potentiality*. Lexington Books, 2015.

Comenius, John Amos. *Orbis Sensualium Pictus*. Translated by Charles Hoole. F. Kirton, 1659.

Coney, John. *Space Is the Place*. Janus Films, 1974. 1 hr., 25 min. https://youtu.be/owCPrIEliZc?si=ADWMvaoYnocqTTPz.

Connolly, William E. *Aspirational Fascism: The Struggle for Multifaceted Democracy Under Trumpism*. University of Minnesota Press, 2017.

Corbin, Henry. *The Man of Light in Iranian Sufism*. Translated by Nancy Pearson. Omega Publications, 1971.

Couillard, Paul. "From No-ing to Knowing, from Naughts to Knots." In *Knowing and Knots: Methodologies and Ecologies in Research-Creation*, edited by Natalie Loveless. University of Alberta Press, 2019.

Cramer, Florian. "What Is 'Post-Digital'?" In *Postdigital Aesthetics: Art, Computation, and Design*, edited by David M. Berry and Michael Dieter. Palgrave Macmillan, 2015.

Critical Computation Bureau. "Editorial — Dialogues on Recursive Colonialism, Speculative Computation, and the Techno-Social." *e-flux Journal* 123, no. 1 (December 2021). https://www.e-flux.com/journal/123/438467/editorial-dialogues-on-recursive-colonialisms-speculative-computation-and-the-techno-social/.

Cronenberg, David, director. *Crimes of the Future*. New Cinema Enterprises, 1970. 1 hr., 10 min. https://youtu.be/Y8PhjIBblPA?si=Dzp1UeuFsIGE8mt4.

Culp, Andrew. *A Guerrilla Guide to Refusal*. University of Minnesota Press, 2022.

Curb Your Enthusiasm. 1999–2024. Created by Larry David. Aired on HBO.

Daston, Lorraine. *Rules: A Short History of What We Live by*. Princeton University Press, 2022.

Daston, Lorraine, and Peter Galison. "The Image of Objectivity." *Representations* 40 (1992): 81–128.

Davis, Erik. *TechGnosis: Myth, Magic and Mysticism in the Age of Information*. North Atlantic Books, 2015.

Debus, Allen G. "Alchemy in an Age of Reason: The Chemical Philosophers of the Early Eighteenth-Century France." In *Hermeticism and the Renaissance: Intellectual History and the Occult in Early Modern Europe*, edited by Ingrid Merkel. Associated University Presses, 1988.

Deleuze, Gilles. *Difference and Repetition*. Translated by Paul Patton, Columbia University Press, 1994.

Deleuze, Gilles. "The Exhausted." In *Essays Critical and Clinical*, translated by Daniel W. Smith and Michael A. Greco. University of Minnesota Press, 1997.

Deleuze, Gilles. *Foucault*. Translated by Séan Hand. University of Minnesota Press, 1988.

Deleuze, Gilles. "Mathesis, Science, and Philosophy." In *Collapse, Volume 3*, edited by Robin Mackay. MIT Press, 2007.

Deleuze, Gilles. *Negotiations, 1972–1990*. Translated by Martin Joughin. Columbia University Press, 1995.

Deleuze, Gilles. "An Unrecognized Precursor to Heidegger: Alfred Jarry." In *Essays Critical and Clinical*, translated by Daniel W. Smith and Michael A. Greco. University of Minnesota Press, 1997.

Deleuze, Gilles, and Félix Guattari. *Anti-Oedipus: Capitalism and Schizophrenia*. Translated by Robert Hurley, Mark Seems, and Helen R. Lane. University of Minnesota Press, 1983.

Deleuze, Gilles, and Félix Guattari. *A Thousand Plateaus: Capitalism and Schizophrenia*. Translated by Brian Massumi. University of Minnesota Press, 1987.

Deleuze, Gilles, and Félix Guattari. *What Is Philosophy?* Translated by Hugh Tomlinson and Graham Burchell. Verso, 1994.

Deleuze, Gilles, and Claire Parnet. *Dialogues II*. Translated by Hugh Tomlinson and Barbara Habberjam. Columbia University Press, 2007.

Di Leo, Jeffrey R. *Dark Academe: Capitalism, Theory, and the Death Drive in Higher Education*. Springer, 2024.

Dixon-Román, Ezekiel, and Ramon Amaro. "Haunting, Blackness, and Algorithmic Thought." *e-flux Journal* 123, no. 1 (December 2021). https://www.e-flux.com/journal/123/437244/haunting-blackness-and-algorithmic-thought/.

Drucker, Johanna. "The Patriarchal Demon, Two Decades Later." In *Pataphysics Unrolled*, edited by Katy L. Price and Michael R. Taylor. Penn State University Press, 2022.

Dufourmantelle, Anne. *In Praise of Risk*. Translated by Steve Miller. Fordham University Press, 2019.

Eamon, William. *Science and the Secrets of Nature: Books of Secrets in Medieval and Early Modern Culture*. Princeton University Press, 1994.

Easterling, Keller. *The Action Is the Form: Victor Hugo's TED Talk*. Strelka Press, 2012.

Eliade, Mircea. *The Forge and the Crucible*. Translated by Stephen Corrin. University of Chicago Press, 1978.

English, Darby. *How to See a Work of Art in Total Darkness*. MIT Press, 2010.

Eshun, Kodwo. *More Brilliant Than the Sun: Adventures in Sonic Fiction*. Quartet Books, 1998.

Evans, Jules. "Accelerationism, Amphetamine Philosophy, and the Death Trip." *Medium*. https://julesevans.medium.com/acceler-ationism-amphetamine-philosophy-and-the-death-trip-bb67ff0 79f8.

Faivre, Antoine. *Access to Western Esotericism*. Translated by Christine Rhone. State University of New York Press, 1994.

Faivre, Antoine. *L'esotérisme*. Presses Universitaires de France, 1992.

Faye, Guillaume. *Archeofuturism: European Visions of the Post-Catastrophic Age*. Arktos Media, 2010.

Ford, Derek R., and Tyson E. Lewis. "On the Freedom to Be Opaque Monsters: Communist Pedagogy, Aesthetics, and the Sublime." *Cultural Politics* 14, no. 1 (2018): 95–108.

Ford, Derek R., Katie Swenson, and Megan Fosher. "From the Knowable and Transparent Individual to the Secret Thought of Individuation: An Anti-Capitalist Postdigital Ecopedagogy." In *Postdigital Ecopedagogies: Genealogies, Contradictions, and Possible Futures*, edited by Petar Jandrić and Derek R. Ford. Springer International Publishing, 2022.

Forshaw, Peter J. "*Oratorium — Auditorium — Laboratorium*: Early Modern Improvisations on Cabala, Music, and Alchemy," *Aries* 10, no. 2 (2010): 169–95.

Fort, Charles. *The Book of the Damned.* Boni and Liveright, 1919.

Franklin, Samuel W. *The Cult of Creativity: A Surprisingly Recent History.* University of Chicago Press, 2023.

Freitas, Elizabeth de. "Nonhuman Findings from the Laboratory of Speculative Sociology." *The Minnesota Review: A Journal of Creative and Critical Writing*, no. 88 (2017): 116–26.

Freud, Sigmund. "Psycho-Analysis and Telepathy." In *Standard Edition of the Complete Psychological Works of Sigmund Freud*, translated by James Strachey. Hogarth Press, 1966.

Friedberg, Anne. *The Virtual Window: From Alberti to Microsoft.* MIT Press, 2006.

Fulcanelli. *Dwellings of the Philosophers, and Hermetic Symbolism in Relationship with Sacred Art and the Esotericism of Grant Work, Volumes One and Two.* Translated by Daniel Bernardo. Sojourner Books, 2021.

Fuller, Matthew, and Eyal Weizman. *Investigative Aesthetics: Conflicts and Commons in the Politics of Truth.* Verso Books, 2021.

Galloway, Alexander R. *Uncomputable: Play and Politics in the Long Digital Age.* Verso, 2021.

Galloway, Alexander R. "What Is Hermeneutic Light?" In *Leper Creativity: Cyclonopedia Symposium*, edited by Ed Keller, Nicola Masciandaro, and Eugene Thacker. Punctum Books, 2012.

Garland-Thomson, Rosemarie. "What Misfitting Makes." *Puncta: Journal of Critical Phenomenology* 7, no. 1 (2024): 5–22.

Garoian, Charles R. *The Prosthetic Pedagogy of Art: Embodied Research and Practice.* State University of New York Press, 2014.

Gaskins, Nettrice R. "Afrofuturism on Web 3.0: Vernacular Cartography and Augmented Space." In *Afrofuturism 2.0: The Rise of Astro-Blackness*, edited by Reynaldo Anderson and Charles E. Jones. Lexington Books, 2015.

Glissant, Édouard. *Poetics of Relation.* Translated by Betsy Wing. University of Michigan Press, 1997.

Goethe, Johann Wolfgang von. *Goethe's Faust.* Translated by Walter Kaufmann. Anchor Books, 1963.

Goldhill, Olivia. "The Neo-Fascist Philosophy That Underpins Both the Alt-Right and Silicon Valley Technophiles." *Quartz*, (June 2017). https://qz.com/1007144/the-neo-fascist-philosophy-that-underpins-both-the-alt-right-and-silicon-valley-technophiles.

Gorsky, William T. *Yeats and Alchemy.* State University of New York Press, 1996.

Grindon, Gavin. "Alchemist of the Revolution: The Affective Materialism of Georges Bataille." *Third Text* 24, no. 3 (May 2010): 305–17.

Guattari, Felix. *Chaosophy: Texts and Interviews 1972–1977.* Edited by Sylvère Lotringer. Translated by David L. Sweet, Jarred Becker, and Taylor Adkins. Semiotext(e), 2009.

Gupta-Nigam, Anirban. "Black Infrastructure: Media and the Trap of Visibility." *Media Fields Journal* no. 11 (2016): 1–11.

Hall, James. *The Artist's Studio: A Cultural History.* WW Norton, 2023.

Han, Byung-Chul. *The Scent of Time: A Philosophical Essay on the Art of Lingering,* translated by Daniel Steuer. Polity Press, 2017.

Hanegraaff, Wouter J. *Esotericism and the Academy: Rejected Knowledge in Western Culture.* Cambridge University Press, 2013.

Haraway, Donna. "Situated Knowledges: The Science Question in Feminism and the Privilege of Partial Perspective." *Feminist Studies* 14, no. 3 (Autumn 1988): 575–99.

Haraway, Donna. *Staying with the Trouble: Making Kin in the Chthulucene.* Duke University Press. 2016

Harney, Stefano, and Fred Moten. "Plantocracy or Communism." In *Propositions for Non-Fascist Living: Tentative and Urgent,* edited by Maria Hlavajova and Wietske Maas. MIT Press, 2019.

Harney, Stefano, and Fred Moten. *The Undercommons: Fugitive Planning and Black Study.* Autonomedia, 2013.

Hartmann, Franz. *The Life and Doctrine of Paracelsus.* John W. Lovell, 1891.

Haskell, Caitlin, and Tere Arcq, eds. *Remedios Varo: Science Fictions.* Art Institute of Chicago, 2023.

Haworth, Christopher. "Post-Punk, Industrial Culture Zines, and the Information Dark Age." *Theory, Culture, & Society* 40, no. 7–8 (2023): 211–35.

Heller-Roazen, Daniel. *Dark Tongues: The Art of Rogues and Riddlers.* Zone Books, 2013.

Hofstadter, Douglas R. *I Am a Strange Loop.* Basic Books, 2007.

Holland, Eugene. "Schizoanalysis, Nomadology, Fascism." In *Deleuze and Politics,* edited by Ian Buchanan and Nicholas Thoburn. Edinburgh University Press, 2008.

Horkheimer, Max, and Theodor Adorno. *Dialectic of Enlightenment: Philosophical Fragments.* Translated by Edmund Jephcott. Stanford University Press, 2002.

Hugill, Andrew, and James Hendler. "Pataphysics and Computing." In *Paraphysics Unrolled,* edited by Katy L. Price and Michael R. Taylor. Penn State University Press, 2022.

Hyland, Peter, and Tyson E. Lewis. *Studious Drift: Movements and Protocols for a Postdigital Education.* University of Minnesota Press, 2022.

Hyland, Peter, and Tyson E. Lewis. "Vestigial Research for Postdigital Pataphysics." In *Postdigital Research: Genealogies, Challenges, and*

Future Perspectives, edited by Petar Jandrić, Alison MacKenzie, and Jeremy Knox. Springer, 2023.

Ibsen, Henrik Johan. *Peer Gynt: A Dramatic Poem*. Translated by Peter Watts. Read Books, 1966.

Jackson, Catherine M. "Analysis and Synthesis in Nineteenth-Century Organic Chemistry." PhD diss., University of London, 2009.

Jameson, Fredric. *Postmodernism, or, the Cultural Logic of Late Capitalism*. Duke University Press, 1991.

Jameson, Fredric. *Valences of the Dialectic*. Verso, 2009.

Jarry, Alfred. *Exploits & Opinions of Dr. Faustroll, Pataphysician*. Translated by Simon Watson Taylor. Exact Change Books, 1996.

Jarry, Alfred. *How to Construct a Time Machine*. Patakosmos Press Open Access, 2013.

Jones, Caitlin. "The Function of the Studio (When the Studio Is a Laptop)." *Artlies 67* (Fall/Winter 2010). https://faa218.files.wordpress.com/2014/08/jones_caitlin.pdf.

Jones, Caroline A. *Machine in the Studio: Constructing the Postwar American Artist*. University of Chicago Press, 1996.

Jones, Ernest. *The Life and Work of Sigmund Freud*. Plunkett Lake Press, 2019.

Jung, Carl G. "The Idea of Redemption in Alchemy." In *The Integration of the Personality*, translated by Stanley Dell. Kegan Paul, Trench, Trubner & Co., 1946.

Kafka, Franz. "The Cares of a Family Man." In *Franz Kafka: The Complete Stories*, edited by Nahum Glatzer and John Updike. Schocken Books, 1995.

Kafka, Franz. *The Castle*. Translated by Mark Harman. Schocken Books, 1998.

Kahn, Richard, and Tyson E. Lewis. *Education out of Bounds: Reimagining Cultural Studies for a Posthuman Age*. Palgrave Macmillan, 2010.

Kalaidjian, Walter. "Occult Surrealism as 'Profane Illumination': Mina Loy, Leonora Carrington, and Ithell Colquhoun." *Journal of Modern Literature* 46, no. 1 (2022): 158–76. https://doi.org/10.2979/jmodelite.46.1.10.

Kaplan, Janet A., and Remedios Varo. *Unexpected Journey: The Art and Life of Remedios Varo*. Abbeville Press, 2000.

Kieffer, Fanny. "The Laboratories of Art and Alchemy at the Uffizi Gallery in Renaissance Florence: Some Material Aspects." In *Laboratories of Art: Alchemy and Art Technology from Antiquity to the 18th Century*, edited by Sven Dupré. Springer, 2014.

Kircher, Athanasius. *Magnes sive de Arte Magnetica*. n.p., 1641.

Kishik, David. *The Manhattan Project: A Theory of The City*. Stanford University Press, 2015.

Klemperer, Victor. *The Language of the Third Reich: LTI — Lingua Tertii Imperii: A Philologist's Notebook*. The Athlone Press, 2000.

Kliën, Michael. *Parliament*. 2013. Social choreography. Originally performed at the Benaki Museum, Athens, Greece.

Kotsko, Adam. *Awkwardness*. Zero Books, 2010.

Kozel, Susan. *Closer: Performance, Technology, Phenomenology*. MIT Press, 2008.

Larkin, Brian. "The Politics and Poetics of Infrastructure." *Annual Review of Anthropology* 42, no. 4 (August 21, 2013): 327–42.

Laruelle, François. "On the Black Universe." In *Dark Nights of the Universe*, edited by Eugene Thacker, Daniel Colucciello Barber, Nicola Masciandaro, and Alexander Galloway. [NAME] Publications, 2013.

Latour, Bruno. *We Have Never Been Modern*. Translated by Catherine Porter. Harvard University Press, 1993.

Latour, Bruno, and Steve Woolgar. *Laboratory Life: The Social Construction of Scientific Facts*. SAGE Publications, 1979.

Le Corbusier. "A Coat of Whitewash; the Law of Ripolin." In *The Decorative Art of Today*, translated by James I, Dunnett. Architectural Press, 1987.

Lee, Key Jo. "Introduction: Defining Perceptual Drift." In *Perceptual Drift: Black Art and an Ethics of Looking*, edited by Key Jo Lee. Yale University Press, 2023.

Lewis, Tyson E. "The American Agitator Goes Digital: Understanding the Affective Role of Agitational Aesthetics in the Online Fascist Recruitment of Youth." *Review of Education, Pedagogy, and Cultural Studies*, DOI: 10.1080/10714413.2023.2291345.

Lewis, Tyson E. "Defining a 'Pataphysical Stance for Arts-Based Research." *Studies in Art Education* 61, no. 3 (September 17, 2020): 220–29.

Lewis, Tyson E. "Fascism's Spatial Imaginary at the Threshold." *Social Text* 42, no. 1 (2024): 27–51.

Lewis, Tyson E. *Inoperative Learning: A Radical Rewriting of Educational Potentialities*. Routledge, 2018.

Lewis, Tyson E. *On Study: Giorgio Agamben and Educational Potentiality*. Routledge, 2013.

Lewis, Tyson E. "The Pataphysics of Inoperativity in the Works of Giorgio Agamben." *Journal of Italian Philosophy* 3 (2020): 139–61.

Lewis, Tyson E. *Walter Benjamin's Antifascist Education: From Riddles to Radio*. State University of New York Press, 2020.

Lewis, Tyson E., and Amelia Kraehe. "Rise: Emergent Strategies for Reclaiming Joy and Agency Against Neofascist and White Supremacist Assaultive Speech." *Journal of Cultural Research in Art Education* 37 (2020): 39–55.

Loveless, Natalie. *How to Make Art at the End of the World: A Manifesto for Research-Creation*. Duke University Press, 2019.

Loveless, Natalie, and Erin Manning. "Research-Creation as Interdisciplinary Praxis." In *Knowing and Knots: Methodologies and Ecologies in Research-Creation*, edited by Natalie Loveless. University of Alberta Press, 2019.

Löwy, Michael. *Fire Alarm: Reading Walter Benjamin's "On the Concept of History."* Translated by Chris Turner. Verso Books, 2016.

Luhmann, Niklas. "Differentiation of Society." *The Canadian Journal of Sociology* 2, no. 1 (1977): 29–53.

Magee, Glenn Alexander. *Hegel and the Hermetic Tradition*. Cornell University Press, 2008.

Magnus, Albertus. "From the *Libellus de Alchimia*." In *The Alchemy Reader*, edited by Stanton J. Linden. Translated by Virginia Heines. Cambridge University Press, 2003.

Maier, Michael. *Furnace and Fugue: A Digital Edition of Michael Maier's* Atalanta fugiens *(1618) with Scholarly Commentary*. Edited by Tara Nummedal and Donna Bilak. University of Virginia Press, 2020. https://furnaceandfugue.org/index.html.

Maly, Ico. "New Right Metapolitics and the Algorithmic Activism of Schild & Vrienden." *Social Media + Society*, (April–June 2019): 1–15.

Manning, Erin. *For a Pragmatics of the Useless*. Duke University Press, 2020.

Manning, Erin. *The Minor Gesture*. Duke University Press, 2016.

Marx, Karl. *Capital: Volume One*. Translated by Ben Fowkes and David Fernbach. Penguin Classics, 1990.

Marx, Karl. "Inaugural Address of the International Workmen's Association." in *MECW: Vol. 20 General Works 1844–1895*. International Publishers, 1985.

Marx, Karl, and Fredrich Engels. *The Communist Manifesto*. Translated by Samuel Moore. Penguin Books, 1967.

Materer, Timothy. *Modernist Alchemy: Poetry and the Occult*. Cornell University Press, 2018.

McCaffery, Steve. *The Darkness of the Present: Poetics, Anachronism, and the Anomaly*. University of Alabama Press, 2012.

McCole, John. *Walter Benjamin and the Antinomies of Tradition*. Cornell University Press, 1993.

Melzer, Arthur M. *Philosophy Between the Lines: The Lost History of Esoteric Writing*. University of Chicago Press, 2014.

Mercer, J. E. *Alchemy: Its Science and Romance*. Society for Promoting Christian Knowledge, 1921.

Merkur, Daniel. *Gnosis: An Esoteric Tradition of Mystical Visions and Unions*. State University of New York Press, 1993.

Merleau-Ponty, Maurice. *Phenomenology of Perception*. Translated by Donald Landes. Routledge, 2012.

Metzger, Richard. "Introduction: Operation Mindwarp." In *Book of Lies: The Disinformation Guide to Magick and the Occult: (Being an Alchemical Formula to Rip a Hole in the Fabric of Reality)*, edited by Richard Metzger. Disinformation, 2008.

Mitchell, Robert. *Experimental Life: Vitalism in Romantic Science and Literature*. Johns Hopkins University Press, 2013.

Mittelmeier, Martin. *Naples 1925: Adorno, Benjamin, and the Summer That Made Critical Theory*. Translated by Schelley Frisch. Yale University Press, 2024.

Moffitt, John F. "Cryptography and Alchemy in the Work of Marcel Duchamp and Walter Arensberg." *Aries* 1 no. 1 (2001): 38–60.

Mohaghegh, Jason Bahbak. *Night: A Philosophy After Dark*. Zone Books, 2019.

Morris, Peter J. T. *The Matter Factory: A History of the Chemistry Laboratory.* Reaktion Books, 2015.

Morrison, Grant. "Pop Magic!" In *Book of Lies: The Disinformation Guide to Magick and the Occult: (Being an Alchemical Formula to Rip a Hole in the Fabric of Reality).* Disinformation, 2008.

Mosse, George. *The Fascist Revolution: Toward a General Theory of Fascism.* University of Wisconsin Press, 2021.

Moten, Fred. "Blackness and Nothingness (Mysticism in the Flesh)." *South Atlantic Quarterly* 112, no. 4 (2013): 737–80.

Moten, Fred. "The Case of Blackness." *Criticism* 50, no. 2 (2008): 177–218. http://www.jstor.org/stable/23128740.

Morton, Timothy. *Dark Ecology: For a Logic of Future Coexistence.* Columbia University Press, 2019.

Newman, William R., and Lawrence M. Principe. *Alchemy. Tried in the Fire: Starkey, Boyle, and the Fate of Helmontian Chymistry.* University of Chicago Press, 2005.

Noble, Judith. "The Wedding of Light and Matter: Alchemy and Magic in the Films of Derek Jarman." In *Visions of Enchantment: Occultism, Magic, and Visual Culture,* edited by Daniel Zamani, Judith Noble, and Merlin Cox. Fulgar Press, 2019.

North, Paul. *The Yield: Kafka's Atheological Reformation.* Stanford University Press, 2015.

Nummedal, Tara. "Sound and Vision: The Alchemical Epistemology of Michael Maier's *Atalanta fugiens*." https://furnaceandfugue.org/essays/nummedal/.

O'Doherty, Brian. *Studio and Cube: On the Relationship Between Where Art Is Made and Where Art Is Displayed.* Princeton Architectural Press, 2007.

Panagia, Davide. "On the Possibilities of a Political Theory of Algorithms." *Political Theory* 49, no. 1 (2021): 109–33. https://doi.org/10.1177/0090591720959853.

Paglen, Trevor. *Blank Spots on the Map: The Dark Geography of the Pentagon's Secret World.* Penguin Publishing Group, 2010.

Pallasmaa, Juhani. *The Eyes of the Skin: Architecture and the Senses.* Wiley, 2007.

Partridge, Christopher. "Occulture Is Ordinary." In *Contemporary Esotericism,* edited by Egil Asprem and Kennet Granholm. Routledge, 2014.

Perloff, Marjorie. *Infrathin: An Experiment in Micropoetics.* University of Chicago Press, 2021.

Pignarre, Phillipe, and Isabelle Stengers. *Capitalist Sorcery: Breaking the Spell.* Palgrave Macmillan, 2011.

Policht, Piotr. "Divas, Junkies and Alchemical Experiments. Anke Kempkes and Bette Stoler in Conversation." *Blok,* 21 November 2018. https://blokmagazine.com/divas-junkies-and-alchemical-experiments-anke-kempkes-and-bette-stoler-in-conversation/.

Price, Katie L. "Introduction." In *Pataphysics Unrolled,* edited by Katie L. Price and Michael R. Taylor. Penn State University Press, 2022.

Principe, Lawrence M. *The Secrets of Alchemy*. University of Chicago Press, 2013.

Plankorius, Zibunspiel. "Sexing the Eel: The Impact of Searching for Eel Gonads on the Later Metapsychology of Sigmund Freud." *Sea Witch Chronicles* 8, no. 10 (1977): 1111–666.

Ramey, Joshua. *The Hermetic Deleuze: Philosophy and Spiritual Ordeal*. Duke University Press, 2012.

Rediker, Marcus. *The Slave Ship: A Human History*. W.W. Norton, 2007.

Ritchie, Matthew. "The Demon in the Diagram: Towards a Theory of Evolving Diagrammatic Embodiment." *Journal of Visual Art Practice* 20, no. 3 (2021): 215–57.

Rogers, Hannah Star. *Art, Science, and the Politics of Knowledge*. MIT Press, 2022.

Röhr, Julius. *Der okkulte Kraftbegriff im Altertum*. Leipzig, 1923.

Roob, Alexander. *Alchemy and Mysticism*. Translated by Shaun Whiteside. Taschen, 2021.

Rousell, David. "Inhuman Forms of Life: On Art as a Problem for Post-Qualitative Research." *International Journal of Qualitative Studies in Education* 32, no. 7 (March 2019): 887–908. 10.1080/09518398.2019.1609123.

Rubenstein, Michael, Bruce Robbins, and Sophia Beal. "Infrastructuralism: An Introduction." *Modern Fiction Studies* 61, no. 4 (Winter 2015): 575–86. https://doi.org/10.1353/mfs.2015.0049.

Saar, Betye. *The View from the Sorcerer's Window*. 1966. Assemblage of color and intaglio etchings and wood window frame with six panes of glass. 30 × 15½ × 1¼ in. Collection of halley k harrisburg and Michael Rosenfeld.

Senholt, Jacob Christiansen. "Radical Politics and Political Esotericism: The Adaptation of Esoteric Discourse Within the Radical Right." In *Contemporary Esotericism*, edited by Egil Asprem and Kaennet Granholm. Routledge, 2014.

Serres, Michel. *The Five Senses: A Philosophy of Mingled Bodies*. Translated by Margaret Sankey and Peter Cowley. Bloomsbury Press, 2022.

Sharp, Andy. *The English Heretic Collection: Ritual Histories, Magickal Geographies*. Repeater Books, 2020.

Shelley, Mary. *Frankenstein*. W.W. Norton & Company, 2021.

Sholem, Gershom. *Alchemy and Kabbalah*. Translated by Klaus Ottmann. Spring Publications, 2006.

Sibnard, Zilfi. *On the Ontological Paradox of Immanuel Kant's Stockings*. Jingernut Press: forthcoming.

Sills, Leslie. *Visions: Stories About Women Artists*. Albert Whitman, 1993.

Simon, Ed. "In Praise of the Paranormal Curiosity of Charles Fort, Patron Saint of Cranks." *Literary Hub*, June 10, 2024 https://lithub.com/in-praise-of-the-paranormal-curiosity-of-charles-fort-patron-saint-of-cranks/.

Smith, Cauleen. "We Already Have What We Need." *MASS MoCa*, n.d., Accessed March 22, 2025, https://massmoca.org/event/

cauleen-smith/#:~:text=These%20spaces%20cannot%20be%20
parceled,is%20how%20it%20should%20be.

Smith, Marq. "DisORIENTations: Travels Through Blackness." In *Bataille's Eye*, edited by Deborah Cullen. The Institute of Cultural Inquiry, 1997.

Smithson, Robert. "A Sedimentation of the Mind: Earth Projects." Essay written in 1968. https://holtsmithsonfoundation.org/sedimentation-mind-earth-projects.

Sohn-Rethel, Alfred. *Intellectual and Manual Labour: A Critique of Epistemology*. Translated by Martin Sohn-Rethel. Haymarket Books, 2021.

Stengers, Isabelle. *The Invention of Modern Science*. Translated by Daniel W. Smith. University of Minnesota Press, 2000.

Stengers, Isabelle. *La Vierge et le neutrino: les scientifiques dans la tourmente*. Les Empêcheurs de penser en rond, 2006.

"Studio Tour with Betye Saar." *Los Angeles Times* https://www.youtube.com/watch?v=4jTZQzQwT_0.

Sun Ra and His Arkestra. *Outer Space Employment Agency*. Total Energy NER3021, 1991, CD.

Sun Ra and His Arkestra. *Space Is the Place*. Blue Thumb Records BTS 41, 1973, LP.

Takala, Pilvi. *The Trainee*. 2008. Installation comprising three videos with a duration of 13:52, PowerPoint presentation, keycard, and letter.

Tavin, Kevin, Mira Kallio-Tavin, and Max Ryynänen. *Art, Excess, and Education: Historical and Discursive Contexts*. Springer International Publishing, 2019.

Thacker, Eugene. *Starry Speculative Corpse: Horror of Philosophy*. Zero Books, 2015.

The Office (American TV series). 2005–2013. Based on *The Office* (British TV series) created by Ricky Gervais and Stephen Merchant. Aired on NBC.

Thornton, Dora. *The Scholar in His Study: Ownership and Experience in Renaissance Italy*. Yale University Press, 1997.

Toland, John. *Clidophorus*. n.p. 1720.

Toscano, Alberto. *Late Fascism: Race, Capitalism and the Politics of Crisis*. Verso Books, 2023.

Trismegistus, Hermes. "The Emerald Tablet of Hermes Trismegistus." *Gurdjieff International Review* 7, no. 1 (Fall 2003). https://www.gurdjieff.org/hermes.htm.

Truman, Sarah, Natalie Loveless, Erin Manning, Natasha Myers, and Stephanie Springgay, "The Intimacies of Doing Research-Creation." In *Knowing and Knots: Methodologies and Ecologies in Research-Creation*, edited by Natalie Loveless. University of Alberta Press, 2019.

Trump, Donald. Fundraising email message to author. October 24, 2021.

Vandigykis, Osmord. "On the Pataphysical Properties of Gibberish in the Lost Poetry of Chasenstrouth Samstron." *False Teeth Bulletin* 1, no. 2 (1996): 19–77.

van Veen, tobias c. "The Armageddon Effect: Afrofuturism and the Chronopolitics of Alien Nation." In *Afrofuturism 2.0: The Rise of Astro-Blackness*, edited by Reynaldo Anderson and Charles E. Jones. Lexington Books, 2015.

Varo, Remedios. *Ciencia inútil, o El alquimista (Useless Science, or The Alchemist)*. 1955. Painting on Masonite. 21 1/4 x 41 1/3 in. (54 x 105 cm), Museo de Arte Moderno, Mexico City, Mexico.

Vickers, Brian. "On the Function of Analogy in the Occult." In *Hermeticism and the Renaissance: Intellectual History and the Occult in Early Modern Europe*, edited by Ingrid Merkel. Associated University Presses. 1988.

Vico, Enea (after Baccio Bandinelli). *Bandinelli's Night Class*. c.a. 1544. Engraving, 12 1/16 x 17 1/4 in. (30.6 x 43.8 cm). The Metropolitan Museum of Art, New York, object no. 17.50.16–135, https://www.metmuseum.org/art/collection/search/358113.

Vidler, Anthony. *The Architectural Uncanny: Essays in the modern Unhomely*. The MIT Press, 1994.

Warlick, M. E. "The Alchemical Androgyne: The Rewards and Pitfalls of a Feminist Approach." In *Visions of Enchantment: Occultism, Magic, and Visual Culture*, edited by Daniel Zamani, Judith Noble, and Merlin Cox. Fulgar Press, 2013.

Warlick, M. E. *Max Ernst and Alchemy: A Magician in Search of Myth*. University of Texas Press, 2001.

Wertheim, Margaret. *The Pearly Gates of Cyberspace: A History of Space from Dante to the Internet*. W.W. Norton & Company, 1999.

wilson, gloria, Joni Boyd Acuff, and Vanessa López. "Conjuring Hands: The Art of Curious Women of Color." *Hypatia* 36, no. 3 (2021): 566–80.

Winters, Joseph Richard. "Between Ecstasy and Abjection: Black Studies and the Excessive Sacred." *Religious Studies Review* 48, no. 1 (2022): 43–47.

Wolin, Richard. "Left Fascism: Georges Bataille and the German Ideology." *Constellations* 2, no. 3 (1996): 397–428. https://doi.org/10.1111/j.1467-8675.1996.tb00037.x.

Wood, Christopher S. "Indoor-Outdoor: The Studio Around 1500." In *Inventions of the Studio: Renaissance to Romanticism*, edited by Mary Pardo and Michael Wayne Cole. University of North Carolina Press, 2005.

Worby, Robert. "Cacophony." In *Music, Electronic Media and Culture*, edited by Simmon Emmerson, translated by Jane. E. Lewin. Ashgate, 2000.

Yergeau, M. Remi. *Authoring Autism: On Rhetoric and Neurological Queerness*. Duke University Press, 2018.

Zosimos. "Of Virtue, Lessons 1–3." In *The Alchemy Reader*, edited by Stanton J. Linden. Cambridge University Press, 2003.

Zuberi, Nabeel. "The Transmolecularization of [Black] Folk: *Space Is the Place*, Sun Ra and Afrofuturism." In *Off the Planet: Music, Sound, and Science Fiction Cinema*, edited by Philip Hayward. John Libbey, 2020.

Index